The Large Hadron Collider

THE LARGE HADRON COLLIDER

The Extraordinary Story of the Higgs Boson and Other Stuff That Will Blow Your Mind

DON LINCOLN

JOHNS HOPKINS UNIVERSITY PRESS Baltimore

Printed in the United States of America on acid-free paper
9 8 7 6 5 4 3 2 1

Johns Hopkins University Press
2715 North Charles Street
Baltimore, Maryland 21218-4363
www.press.jhu.edu

Library of Congress Cataloging-in-Publication Data

Lincoln, Don, author.
The large hadron collider: the extraordinary story of the Higgs boson and other stuff that will blow your mind / by Don Lincoln.
pages cm
Includes bibliographical references and index.
ISBN 978-1-4214-1351-8 (hardcover: alk. paper) — ISBN 1-4214-1351-5 (hardcover: alk. paper) — ISBN 978-1-4214-1432-4 (electronic) — ISBN 1-4214-1432-5 (electronic) 1. Higgs bosons. 2. Large Hadron Collider (France and Switzerland) I. Title.
QC793.5.B62L557 2014
539.7'360944—dc23 2013040921

A catalog record for this book is available from the British Library.

Special discounts are available for bulk purchases of this book. For more information, please contact Special Sales at 410-516-6936 or specialsales@press.jhu.edu.

Johns Hopkins University Press uses environmentally friendly book materials, including recycled text paper that is composed of at least 30 percent post-consumer waste, whenever possible.

Contents

Preface

September 2008 was an exciting time. In front of a billion sets of eyes, the CERN Large Hadron Collider was turned on. Beams were circulated and champagne flowed. The future looked bright indeed.

The drama intensified when, just a handful of days later, a design flaw in the LHC's electrical protection system led to an uncontrolled electric arc that pierced the vessel holding tons of liquid helium. The helium vented with so much force that it pushed magnets weighing 35 tons off their mounts and fractured the concrete floor into which they had been anchored. Over the course of a few minutes, more than 2 miles (3.2 kilometers) of the LHC tunnel was filled only with cold helium gas, with all trace of ordinary air pushed out of the way. Hours later, movies of the inside tunnel taken by the CERN fire brigade showed thick blocks of ice hanging off the walls and equipment.

The damage was extensive, and the design flaw was still part of the rest of the accelerator. Repairs were in order. After tens of millions of dollars and over a year of work, the LHC resumed operations at reduced beam energy and brightness but still at levels high enough to exceed previous records. And, as they say, the rest is history.

In 2012, experiments at the LHC announced the discovery of the Higgs boson, a particle that had eluded detection for nearly half a century. The story arc of emotional highs, to despair, to triumph was complete. Hundreds of papers, covering the vast range of particle physics, had been written and published on data from the collider. Over a thousand students had worked on or achieved their doctorates. By any standard, the LHC was a success.

When I wrote my earlier book *The Quantum Frontier*, I sent it to the publisher in the first days of 2008, and thus the book was written even before the LHC turned on. It told the tale of the dreams of physicists but nothing of our accomplishments. That story was in the future.

But now that tale has been told, unfolded on that most public of stages. Unlike many other important discoveries in particle physics, the world has watched. More than once, announcements of progress in the story of the pursuit of the Higgs boson got above-the-fold coverage in international newspapers and on the main websites of online news sites. People cared.

So this brings us to the book you're holding. In the years since *The Quantum Frontier* was written, a lot has happened. Many want to know the inside scoop—what it is like to be at the very forefront of research. They want someone to tell them how it happened. They need to hear how it all went down, told by someone who was there—someone . . . well . . . like me. I was there. I sat in the meetings, heard the arguments, felt the elation and the heartbreaks. It has been a glorious time, and the story is one worth telling.

So here it is. This book necessarily has some overlap with the earlier book. The physics we have long known and the accelerators and detectors haven't changed very much. For those topics for which I couldn't think of a better way to say it, I have borrowed from my previous text.

But much of the book is entirely new. Obviously the saga of the startup and the discoveries weren't present before. The Higgs boson now warrants an entirely separate chapter, which explains the theory and the exciting search (including several different analogies, each designed to illuminate a different facet of this fascinating particle). With the discovery of the Higgs boson, the next burning question at the LHC is why its mass is so low. Nobody knows the answer to that question, but it is definitely the next hot topic for LHC physicists, so I explore it in some detail. In fact, in this new book, all of the possible new physics I describe and that we might encounter are tied to the unexplained enigma of the low mass of the Higgs boson. One possible new physical phenomenon we might encounter and could answer that question is extra dimensions of space, which (if true) could actually solve many mysteries. I didn't include a discussion of this subject in *The Quantum Frontier*, but you'll find it here.

Perhaps the most important thing is that we now have operational experience with the LHC. We have a much better idea of what the future might bring us. When the accelerator turns back on and collisions resume, we have a good idea of what to expect and the rate at which we will collect data. The LHC will dominate the energy research frontier for at least another 15 years, and in this book I sketch out what our plans for the future are. Of course, a discovery will inevitably alter those plans, but once you have read this book, you'll know what I know now.

The Large Hadron Collider is the biggest scientific facility ever built, requiring nearly ten thousand physicists and technical professionals to operate it and to understand the data. It was two full decades in planning and building, and it is an international endeavor, supported by most of the world's leading economies. It is likely if you're reading this that universities and laboratories in your own country are participating and, through a tiny fraction of your annual taxes, you have personally contributed to this unparalleled scientific mission. You should know what you are getting for your money. And the answer is simple. Physicists are doing nothing less than coming to understand the deepest rules that govern matter and energy, space and time. So read on and share the enthusiasm of people asking and answering timeless questions. This is a very exciting moment in history.

Acknowledgments

First and foremost I'd like to thank the physicists, engineers, computing professionals, technicians, and other support staff who had the vision and determination to make the Large Hadron Collider and its associated detectors a reality. The LHC is one of the most complex scientific endeavors ever attempted, and I have the greatest respect for a group of people who can make it all work. As more scientific results come in, and certain people become known as the "voice of the LHC," we should never forget the teams that designed and built this equipment. Without them, those voices would be forever mute.

As always, Dan Claes has contributed custom-drawn figures for me. I am always grateful for help in creating these books, but Dan makes it easy. He combines the credentials of a professional physics researcher with great drawing skills. All I had to do was give him a sketchy idea of what I needed and then forget the request. Eventually a perfect figure would appear, frequently better than I had imagined. I'd also like to thank Barry Panas, Flip Tanedo, and Jeffery Mitchell for various computer-generated figures.

The following people read early versions of the text: Sowmya Anjur, Lee Blakley, Peter Dong, Clare Leahy, and David Wang. All of them made invaluable suggestions to clarify the manuscript.

While the earlier readers gave advice on the readability of the book, I also needed help with technical fact checking. For that, I asked my colleagues to double check facts and figures for which they had particular expertise. Marzio Nessi checked the ATLAS description, while David Barney checked the CMS section. Yves Schutz and Roger Forty read the ALICE and LHCb portions, respectively, and Michael Koratzinos vetted the accelerator description. James Gilles, is the official media contact of CERN, and he identified these experts, each with a talent for public communication and a willingness to help out.

Tim Tait has always been very generous with his time to fact check my descriptions of theoretical physics for all my particle physics popularizations. His careful attention to detail was very helpful in ensuring my analogies didn't introduce any unforgiveable errors. In addition, Wally Greenberg and Flip Tanedo provided independent input on the theoretical passages. I remain in their debt.

Of course, there no doubt remain some errors in the text, no matter how many readers have proofread it. In previous books I have assigned the blame to a childhood friend, but I think he's done his penance. Thus I'm afraid I must accept responsibility for any residual errors.

I absolutely must thank the staff at Johns Hopkins University Press, starting with executive editor Vincent Burke. He agreed that my original LHC book *The Quantum Frontier* (also published by JHUP) should be updated for when the LHC began operating at full capability. Michele Callaghan did a wonderful job in editing the original manuscript, polishing off the many rough edges. I should also like to thank the typesetting and production staffs for their role in making this book a reality. Last, but not least, Kathy Alexander has done a great job publicizing the book. An author is responsible for writing a manuscript, but it takes a village to make a book.

And finally, I must thank my family for putting up with my absences during this process.

The Large Hadron Collider

1

BEGINNINGS AND BUILDING BLOCKS

Orderly and chaotic. Cosmically vast and unfathomably tiny. Existing for billions of years in the past and countless eons in the future. Frigidly cold and so blisteringly hot that matter and energy lose their identity, converting back and forth, one to the other, as Einstein predicted over a century ago. As alien as matter swirling down to certain doom in a ravenous black hole and as familiar as a stroll in the park.

This is the universe in which we live.

The size and age of the universe and the dizzying array of phenomena we encounter makes trying to understand it all a challenging and even a daunting goal. What possible similarities are there between the explosion of a supernova that it is so cataclysmic and so bright that it can be seen halfway across the cosmos and a tranquil butterfly dancing around a beautiful and aromatic flower? Is there anything in common with the gentle waves on a pond on a calm day and the maximum speed a skydiver achieves as she plummets to Earth?

We know that there are rules to the universe. But what do we know about them? How many are there? Or do these rules spring from other rules? If you've ever spoken to a three-year-old, you might have gotten into the game of "Why?" The toddler might ask a question like "Why do kittens have fur?" When you reply that it is to keep them warm, the child might follow with "Why do they need to be warm?" You might say that it's because they would feel uncomfortable if they were cold. This response is often another question, perhaps asking why a kitten would be uncomfortable if it was cold. And so it goes.

Depending on your patience and child-rearing skills, you might abruptly cut off the youngster's curiosity with a "Don't ask so many questions" or "Go play with your sister." Or you might answer each "Why?" as best as you could, all the while knowing that your reward will be yet another "Why?" Eventually you'll get to the limit of your knowledge and have to admit that you don't know. This little game of questions does illuminate an important point, which is that many rules originate from even deeper causes. Equally important is the fact that eventually there will be questions for which no answer is known.

While the "Why?" game is sometimes the province of the young, there are grownups who never tire of it. No matter their age, whether 20, 30, or 90, they continue to ask the same kinds of questions, probing each answer with an investigation into the deeper cause. While curiosity respects no education, certainly among the most professionally curious today are scientists.

Individual scientists each have their particular realm in which they are the most curious, with biologists interested in living matter, meteorologists interested in the weather, and so on. But while a meteorologist can predict the weather tomorrow, a deeper understanding of precipitation or wind can only be gained by better understanding the nature of air or how water evaporates. This leads to questions about the molecular makeup of nitrogen, oxygen, and water and how they interact. The answer to these chemical questions can explain the deeper origins of the phenomena studied by the meteorologist, but a curious person might then wonder why the molecules have the properties they do. This "Why?" question turns our attention to the atomic realm. Ask a couple of more "Whys?" and you inevitably end up rummaging around in the territory studied by the particle physicist or cosmologist.

The reason I say this is that, as we delve deeper and deeper, we find that all matter is made of atoms, which is made of even smaller parts. The smallest building blocks we've uncovered thus far are the quarks and leptons that we will learn about chapter 2.

The building blocks are only half of the story. If these components didn't interact, the universe would be a very different place. Individual sand grains do not interact a great deal, with the result being a sand dune or a beach that constantly shifts with the winds and the tide. However, if you add the binding agent of cement, then sand can be fashioned into soaring buildings or towering dams. The fact that the sand can be made into cohesive forms makes it possible for it to be used as a durable construction material in ways that loose sand cannot. The binding greatly changes the physical properties of sand.

Similarly, the quarks and leptons interact with each other to form the atoms that compose our world. Four forces tie everything together, first binding quarks into protons and neutrons with the strong nuclear force and also protons and neutrons into atomic nuclei. These particles lead to atoms formed from atomic nuclei and electrons bound together by the electromagnetic force. Our sun and world were created because the force of gravity drew together bits of matter and formed planets and stars. Our sun burns, in part, by effects of the fourth force, the weak nuclear force. The weak nuclear force converts the hydrogen nuclei in the sun into deuterium, starting the chain of reactions that gently warms you on a mild spring day.

The fact that the familiar universe is composed of two classes of building blocks and a mere four forces is a pretty amazing observation: the simple can generate the complex. Of course, a complete understanding is a lot more complicated than that; for instance, you should not believe that a detailed study of the quarks, leptons, and forces can easily explain the hunting habits of an ordinary housecat; it is for this reason that other sciences remain useful. These disciplines ignore the quarks and leptons and rather simply take as their starting point more complex structures, say a neurologist taking for granted the existence of cells and their chemical interactions or a psychologist starting with the human brain. But the most intricate structures you can imagine have a clear, uninterrupted chain of origins back down to the tiniest microcosm.

If we believe we understand the nature of the smallest building blocks we've discovered (while realizing that there may well be smaller and yet-undiscovered components), we can just as well ask ourselves about the origins of these elemental bits. Like all ultimate origins, their roots can be found by studying the beginning of the universe.

All evidence suggests that about 13.8 billion years ago, the universe began in a cataclysmic explosion called the big bang. Since then, the universe has been expanding and cooling, leading to the familiar world in which we live. Clearly this means that the universe was once smaller and hotter. And hotter means higher energy.

Thus, if we wish to experiment with the deepest and most fundamental nature of the universe's building blocks, as well as their origins, the only way to do that is to attempt to re-create the primordial cauldron in which the cosmos was brewed. To do that, we must somehow heat matter to the highest temperatures our technology allows.

The history of heating begins with the discovery of fire and the mastery of the forge. However, such technologies have long been put aside by those trying to achieve the highest temperatures. Even the heat of the hottest torch pales in comparison to the temperatures that scientists can achieve by accelerating subatomic particles to nearly the speed of light and slamming them together. The fieriest conditions we can achieve in our equipment can exceed those at the center of the sun by more than a hundred-thousand fold. Further understanding the origins of the universe and its fundamental building blocks will require hotter temperatures still.

This book describes mankind's latest attempt to probe the frontiers of our ignorance. Located just west of Geneva, Switzerland, a new particle accelerator is operating. This accelerator is called the Large Hadron Collider, or LHC. It accelerates two beams of protons to the prodigious energy of up to 7 trillion electron volts (TeV) each and collides them at the center of four specially designed detectors that record the details of the collisions. A trillion of anything sounds huge, but on human scales this amount of energy is modest—about as much as two athletic mosquitos flying into each other at top speed. The "prodigious" part comes when you realize that the energy is concentrated in a tiny volume, resulting in those unimaginable temperatures that dwarf those at the sun's core.

Through studies of these collisions, scientists are able to reproduce the conditions of the universe a scant trillionth of a second after it began. This allows a careful study of the rules that governed our beginning and the important factors that have led to the world in which we live. Combined with studies performed in earlier decades, these data provide an excellent understanding of the history of our cosmos, from the earliest times probed by the LHC through the formation of atoms and molecules, stars, and planets. While there are surely mysteries still to explore, it is clearly an epic achievement to be able to understand the history of our universe from a trillionth of a second after the beginning through its current age of 13.8 billion years.

In the pages that follow, I detail our current understanding of the rules that govern matter and energy. I also describe the achievements and challenges of the Large Hadron Collider. The LHC began in earnest in 2010 and has been successful beyond any reasonable expectation. After a period of refurbishment in 2013 and 2014, the LHC should resume operations at an energy that could be as much as double what it delivered in its first years.

The LHC is the most powerful scientific instrument ever devised to study the most fundamental questions of the universe, such as how it evolved and why it has to be the way it is. The scientists studying these data are following the path blazed by generations of earlier intellectual pioneers, each one writing his or her own page in the book of knowledge, a book whose first pages were penned at least 2,500 years ago but whose origins predated writing by millennia.

So I invite you to join me as I tell you this awe-inspiring tale and read to you from the latest chapter in that magnificent book. Let us begin.

2

STUFF WE ALREADY KNOW

On their craggy peninsula in the Aegean Sea, the early Greek philosophers debated long and hard about whether the natural state of matter was resting or moving and whether a smallest particle of matter existed. While many before them surely mulled over the nature of the world they saw around them, the ancient Greeks recorded their thoughts so that others, separated by both space and time, could appreciate and build upon their ideas and debates. In the recording, they tacitly laid claim to the origins of fundamental science.

Much has been written of these long-dead thinkers, but this book is not concerned with their specific thoughts. After all, their ideas were only generally correct and downright wrong in many specifics. However, we *are* concerned with their intellectual legacy.

While the early Greeks may be credited with the start of the journey toward our current understanding of the universe, this understanding has been clarified in the intervening centuries. Our mastery of the natural world includes curing deadly diseases, learning to fly, and the first steps toward re-creating the hot, all-consuming nuclear flame that fuels the sun.

In 1803, the British poet William Blake wrote "The Auguries of Innocence," which began

> To see a world in a grain of sand
> And a heaven in a wild flower,
> Hold infinity in the palm of your hand
> And eternity in an hour.

Seeing the world in a grain of sand is surely a metaphor, but it is not without an element of truth. By considering a single grain of sand and attempting to understand all of its fundamental pieces, one can learn a great deal about the laws that govern the greater universe. For instance, is there a smallest bit of sand? Under a microscope, sand looks a lot like a very small rock. If we crush the grain of sand, we are left with what appears to be even smaller rocks. If we crush those, do we have an infinite chain of ever-smaller rocks?

Asking this question for all the disparate substances of the world—rocks, water, air, food, and so on—led scientists to realize that all the matter of the universe could be created by combing different amounts of some one hundred substances. We call these primordial substances *elements,* and some of their names are familiar from chemistry class. Combine hydrogen and oxygen and get water. Combine sodium and chlorine and get salt. In fact, if you mix the right elements in just the right way, you can make absolutely anything.

So one might ask whether these elements could be subdivided into individual units; that is to say, "Is there a smallest unit of oxygen?" And, indeed it happened to be true, with each element having a smallest piece. We call these smallest pieces *atoms,* and it turned out that the atoms of each element were distinct. If you want to have a basic mental picture of the elements and atoms, think of an old-style toy store that sells children's marbles. One bin contains yellow marbles, while another bin contains big red marbles, and yet another contains tiny green ones. Each bin contains marbles of a distinct size and color. All the marbles in each bin are identical, and no two bins have marbles identical to any other bin.

So too it is with the elements and atoms. All of the atoms of an element are identical, and the atoms of different elements are different. And anything on Earth can be made by arranging the right combination of atoms in the right configuration. While the details of how you do the mixing are quite complex, one can learn a lot of chemistry just by using this simple analogy of marbles. Figure 2.1 lists the elements we've found thus far. The chart is called the Periodic Table of the Elements, or just the periodic table for short. Each block denotes a particular element. Columns indicate elements that react similarly when combined with other elements.

By the early 1900s physicists came to realize that atoms could themselves be broken down into smaller components. By 1932, physicists discovered that all atoms could be assembled by the right mix of three even smaller particles, called protons, neutrons, and electrons. All protons were identical, as were neutrons and electrons. The discovery that with a hundred different kinds of atoms one could make anything in the cosmos was an astounding simplification. But we now knew that the elements themselves could be made from the right combinations of these three smaller ingredients. For example, a hydrogen atom could be made from one proton and one electron. Helium atoms required two protons, two neutrons, and two electrons. The patterns for the other atoms were eventually worked out.

As soon as it became clear that atoms could be constructed of smaller particles, there naturally was interest in trying to figure out how the particles were arranged inside the atom. For instance, were the protons, neutrons, and electrons

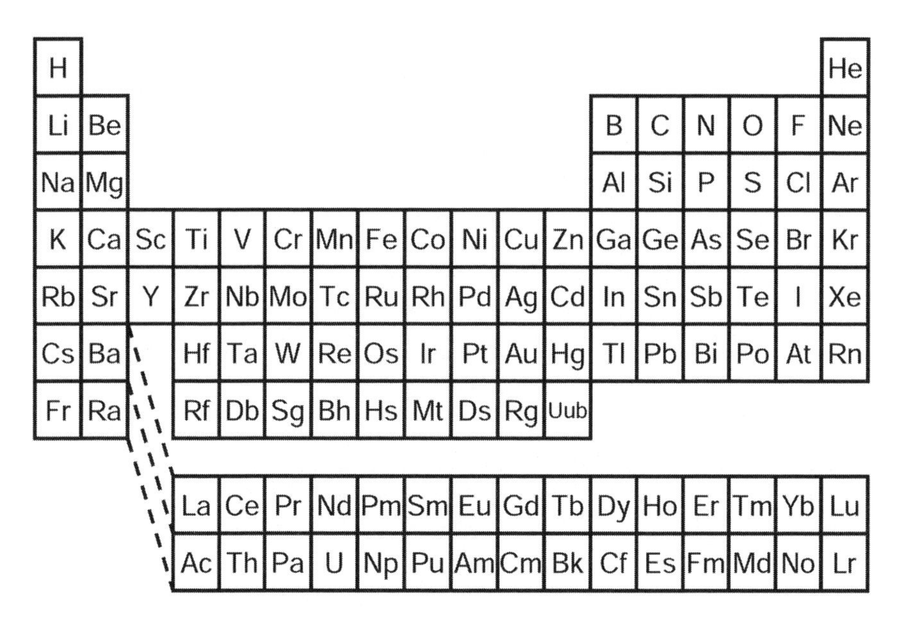

Figure 2.1. When the chemical periodic table of elements was first proposed in the 1860s, it was a radical innovation. Instead of countless unrelated substances, it became clear that all the matter of the universe could be explained as various mixtures of a finite number of elements.

all clumped together in a tapioca-like mass? Or perhaps they were lined up like beads on a string. Logic really couldn't guide us to decide what an atom looked like. For that we needed experiments.

It was Ernest Rutherford who figured out the rough structure of the atom. He found that the atom was rather like a little solar system. From his work and that of others, it was shown that each atom had equal numbers of electrons and protons. The protons were all clumped together with the neutrons in a tiny ball that is called the *nucleus* of the atom. The electrons could swirl around the nucleus at a relatively great distance. Following the solar system analogy, the nucleus was equivalent to the sun, while the electrons were more like the planets. The protons were found to have positive electrical charge and the electrons had precisely the same amount of charge, but negative. Exactly why this should be the case is not known even today. The neutrons were electrically neutral. Each atom had equal numbers of electrons and protons. The number of neutrons doesn't follow such simple rules. With the exception of hydrogen, the number of neutrons in an atom is similar to the number of protons but usually a bit higher.

While the protons and electrons have equal electrical charge (although opposite in sign), they have radically different mass. The proton is about two

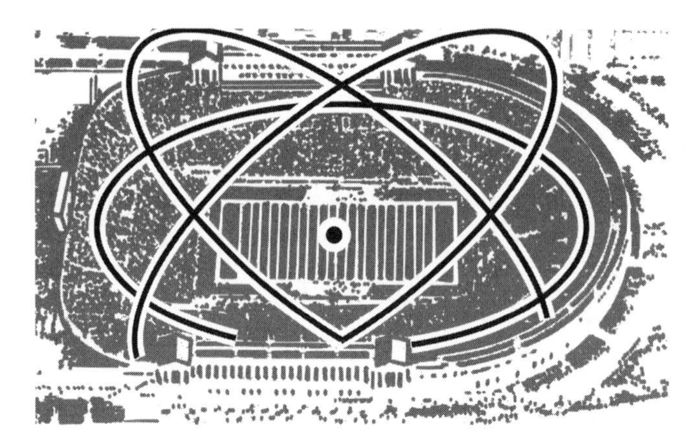

Figure 2.2. Atoms are mostly empty space. If the electron cloud of an atom were blown up to the size of a football stadium, the protons and neutrons would be about the size of the letter "o" on this page. Figure courtesy of Dan Claes.

thousand times more massive than the electron. (Note that when physicists use the word "massive," it means that something has a mass and is contrasted with "massless," which means having no mass. Anything with any mass at all is called massive although, in the particle realm, massive things are actually incredibly tiny.) The neutron's mass is a smidge larger than the proton's mass. This disparity in the masses of the atom's components means that something like 99.95% of the mass of an atom is in the nucleus.

While the protons and neutrons inhabit the nucleus of the atom, with the electrons swirling around at a relatively large distance, this doesn't really give us an idea of the size of an atom. Atoms are really, really, really small. In fact, if you decided to line up atoms "edge to edge," it would take 250 million to make up a single inch (2.5 centimeters). Even after one realizes just how small the atom is, not even that really gives the full picture. The atom consists of mostly empty space, with the diameter of the nucleus of the atom being about ten thousand times smaller than the atom itself.

Let's use an analogy to get an idea of just how mind-bogglingly empty an atom is. Think about a carbon atom, one of the building blocks of life. A carbon atom consists of six protons and six neutrons in the nucleus, with six electrons swirling around the nucleus at a great distance. Imagine we blew up each proton or neutron to be a sphere the size of a printed "o" on this page (figure 2.2). We could think of the nucleus as six of these BB-sized spheres (the protons) and six spheres (the neutrons) all clumped together. We could imagine the protons and neutrons as each having a different color to distinguish them in our mental picture, but that's just artistic license to remind us that they are different. Color doesn't have a meaning at the subatomic level. Why don't we further put this analogy nucleus at the 50 yard line of Soldier Field, home of the Chicago Bears football team? If we did this, the rest of the atom would consist of six electrons, each much smaller than a printed period on this page, swirling like frenzied bees in a sphere the size of the football stadium. The atom is almost entirely empty space. Even so, these tiny and empty atoms of a hundred different elements, each consisting of only protons, neutrons, and electrons, make up everything in the entire universe.

Quarks

You'd think that scientists would celebrate the realization that, with three tiny particles, they could explain the entire universe and then they'd leave well enough alone. But we physicists are a curious lot, and the scientists of the time kept poking at the question. In the 1940s and 1950s, physicists studied the data coming from their new toys, such as the so-called atom smashers, and from cosmic rays, which seemed to be raining down on Earth from space itself. They discovered particles in their data that did not fit neatly into the proton, neutron, electron, or atom classification scheme. In fact, they found nearly a hundred different particles that seemed have similarities with the primordial protons, neutrons, and electrons, although each newly discovered particle had its own identifying quirk. These particles were given names like pions, kaons, lambdas, and Vs, although there were tons more. It was a veritable soup of letters and Greek symbols. Scientists scratched their heads.

The scratching went on for quite a few years until 1964, when a very clever proposal was made. Maybe the primordial protons and neutrons weren't that fundamental after all. Perhaps they themselves were made of even smaller objects. These objects have come to be called quarks (pronounced so it rhymes with "forks" by Americans, although some of my British colleagues pronounce the word so it rhymes with "barks"), after an inconsequential line from James Joyce's *Finnegans Wake* ("Three quarks for Muster Mark!"). Unlike earlier choices for the names of fundamental particles (both the words atom and proton have Greek antecedents: *atomos* for "not able to be cut" and *protos* for "first"), the word "quark" has no such classical inspiration and fits well with modern physics' tradition of whimsical names.

While originally only three quarks were proposed, we now know of six. Their names are: up and down, charm and strange, and top and bottom. The names don't really have any deeper meaning. Of all the quarks, two are by far the most prevalent: the up and down quarks. These two make up the proton (consisting of two ups and one down) and neutron (one up and two downs). The others are necessary to fully explain the plethora of particles discovered in

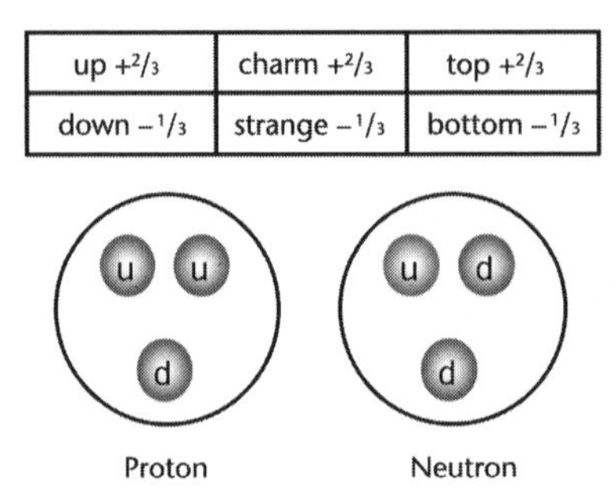

Figure 2.3. ***Top,*** **modern research has identified six types of quarks: up, down, charm, strange, top, and bottom. The fraction in the figure denotes the charge of that particular quark, in units in which the charge of the proton is +1.** ***Bottom,*** **protons and neutrons, along with electrons, make of the atoms of all ordinary matter. They are composed of three quarks: proton (up, up, down) and neutron (up, down, down).**

particle accelerators (the pions, kaons, lambdas, and Vs listed above, as well as many others). Figure 2.3 lists the six quarks and shows how they make up the proton and neutron.

The first three quarks proposed were the up, down, and strange quarks. The names "up" and "down" come from an older theory of the nucleus in which the protons and neutrons were treated as essentially the same thing. Up and down had a technical meaning but can be thought of as being similar to the two sides of a single coin. The language of this older theory was carried over to the quarks. The name "strange" also is a historical holdover. Some of the particles discovered in the early accelerator and cosmic ray experiments acted oddly and people said, "Huh! That's strange . . . " It turned out that their unusual behavior was related to the fact that they contained a strange quark within them, so the name migrated from the strange particles to the quark. Consequently, it's a bit tricky to say when the up, down, and strange quarks were discovered, since we saw them in the first six or so decades of the twentieth century. It was only in 1964 that we recognized them for what they were and in the 1970s that they were physically observed. The up quark has an electrical charge two-thirds that of a proton (+⅔), while both the down and strange quarks have only one-third the charge of the proton but with the opposite sign (−⅓). It seemed odd to have two quarks with −⅓ charge and only one with +⅔ charge, but that was how the theory was initially formulated.

The charm quark supposedly gets its name because somebody said, "Wouldn't it be charming if there were a fourth quark, this one with +⅔ charge like the up quark?" It's hard to tell if this is true or physics folklore, but the charm quark was simultaneously discovered in 1974 by two experiments, each located on one of America's coasts, at the Brookhaven National Laboratory in New York State and the Stanford Linear Accelerator Laboratory in California. The bottom quark was discovered in 1977 at Fermilab in Illinois, as was the top quark in 1995. I was one of the discoverers of the top quark as part of two competing teams of physicists, each comprising some 500 scientists. The names "top" and "bottom" have no real meaning, although for a while the words "truth" and "beauty" vied for the honor of naming the two heaviest quarks. The use of these two alternative terms has declined over the past decade and is now pretty rare. That's kind of a shame, as I liked to tell people who came to my public lectures that I was "searching for truth . . ."

With the introduction of quarks, we are approaching one boundary of the current frontier of knowledge. Thus it is important to pause and learn something of their nature. As best as we currently know, quarks are one class of fundamental particles. There are other types, and we'll mention them soon. *Fundamental* in a physics context means that to the best of our knowledge quarks have no size and contain nothing smaller within them (i.e., they have no internal structure). Basically, in the journey into the heart of matter, we are made of molecules, which are in turn made of atoms. Atoms are made of protons, neutrons, and electrons; and protons and neutrons are made of quarks. But when we get to quarks, it's the end of the road. That's it. Quarks are as small as things get, or at least so goes current thinking. Figure 2.4 illustrates the various levels of the microworld for which we have some knowledge.

Naturally, it may well be true that quarks actually are made of even smaller things. In fact, such a possibility is just one of the exciting questions on which the LHC might shed some light. We'll talk about this possibility in chapter 7, but for the moment, we'll concentrate on what we know about quarks.

Of the six types of quarks, only two of them (up and down) are needed to make up protons and neutrons and consequently are stable, which means they don't decay, or change into other particles, under ordinary circumstances. The other four quark types (charm, strange, top, and bottom) all have very short lifetimes, existing for just a fraction of a second before decaying quickly into the more mundane up and down quarks.

Quarks have special rules governing how they can combine. As we have seen, it takes three quarks to make a proton (two ups and a down) or a neutron (two downs and an up). We now know that this requirement of exactly three quarks extends to any particle of the class that includes protons and neutrons. Another class of particles can be made by combining one matter quark with one antimatter quark, but these are only mentioned here for completeness. Antimatter is a concept that will be described toward the end of this chapter and in more detail in chapter 7.

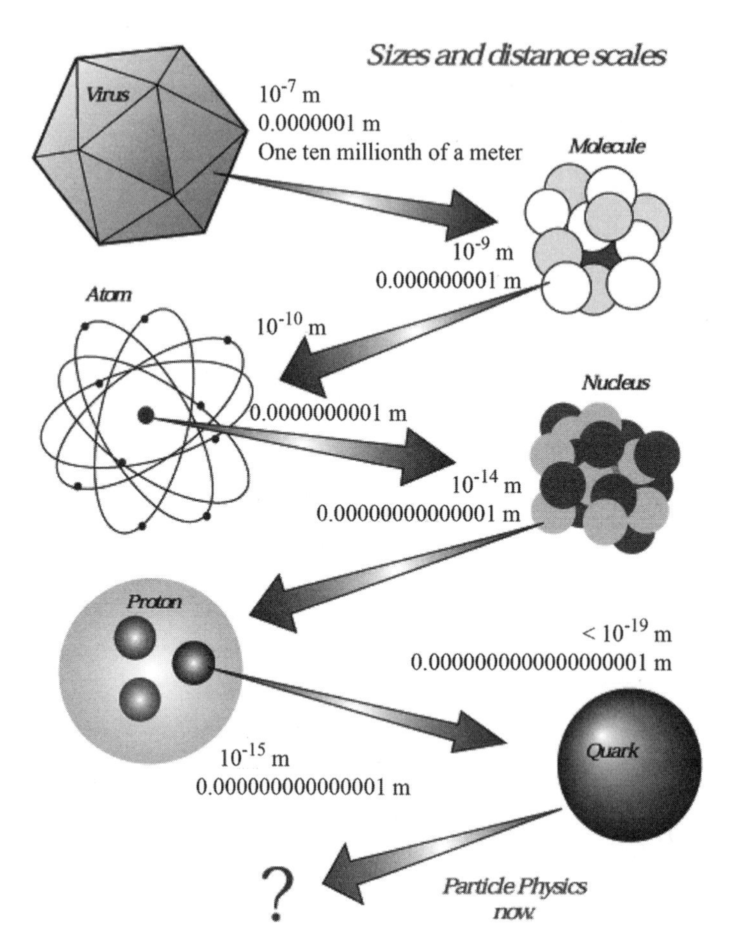

Figure 2.4. The history of particle physics is littered with the remains of particles that were proposed to be the ultimate building blocks of matter. While objects like atoms and protons were once thought to be the smallest particle of matter, we now know this isn't true. It is likely that quarks, which are the smallest particle currently known, will ultimately be found to consist of smaller particles still. Figure courtesy of Fermilab.

In order to appreciate quarks, we have to peek ahead to the idea of forces. While most people have at least a passing familiarity with gravity and electricity, far fewer people are aware that there are two other forces, called the *strong* and *weak nuclear forces*. These two, whose names we shorten to simply the strong and weak forces, only have an appreciable effect in the nucleus of an atom, with the strong force holding the nucleus together and the weak force governing some types of radioactive decay.

The strong force plays an important role in how quarks behave. Originally, the strong force was understood only as that which holds protons and neutrons together in the nucleus of the atom. While there were earlier theories on how this force worked, the picture was greatly simplified by the realization that quarks inhabit the protons and neutrons. It turns out that, just as quarks have electrical charge and consequently feel the electrical force, they also have a new type of charge that governs the strong force. This strong force keeps the quarks in the protons and neutrons and holds the nucleus of an atom together.

This new type of charge is properly called the strong nuclear force charge, but we refer to it as *color*. In contrast to electromagnetism and gravity, which each have only one type of charge (electric charge and mass), the strong nuclear force has three colors. The word in this context bears absolutely no relationship to the ordinary meaning of the word. We use it simply because of a convenient analogy. If you take red, blue, and green lights and simultaneously shine them on a wall, the result that you see is white light, which one might colloquially call no color at all. Similarly, while individual quarks have color charge, if you take them three at a time and put them in a proton, the proton has no net color charge. So we say that quarks can have three types of strong nuclear charge: "red," "blue," and "green." Further, it is true that each proton and neutron always contain three quarks, each with a different color. It is not possible to have a proton with two or three red quarks, because protons have no net color charge, and only by combining red, blue, and green can one get white.

In figure 2.5, we see that the color (strong charge) is unrelated to the quark type. For example, we can look at the down quark and see that it can have any color. To make a proton, all that is required is two up quarks and one down quark, each of which must randomly have one of the three strong force colors (red, green, or blue). Maybe this is easiest to see if we compare it to positive and negative numbers. For numbers, $(+1) + (-1) = 0$. For quarks, we say red + blue + green = 0 (or, equivalently, white).

In discussing color, we are led to another interesting feature of quarks. No quark has ever been directly observed. This doesn't mean that there is no evidence for quarks; indeed the evidence for their existence is simply overwhelming. But it turns out to be impossible to pull a quark out of the atom and study it. Unlike a sandbox, from which you can pull out a single grain of sand to look at, quarks are locked firmly in their respective protons and neutrons. This fact is a consequence of how the strong force acts. The strong force is similar to a spring, in that as you stretch a spring, it gets harder and harder to stretch it more. Contrast this to the electric or magnetic forces, which get weaker as two charged particles are pulled apart. As an example, think of two magnets, which get harder and harder to keep apart (or push together) the closer you bring them to one another. Conversely, when the magnets are far apart from one another, they

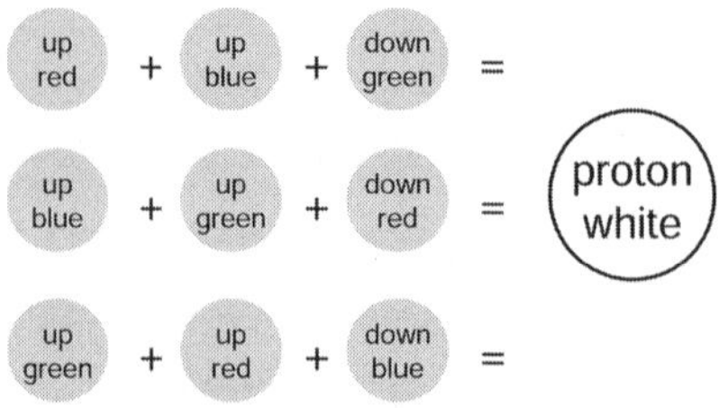

up red + up blue + down green =

up blue + up green + down red = proton white

up green + up red + down blue =

Figure 2.5. If you take three lights that are red, green, and blue and shine them on a white wall, the resultant light will be white. The situation is similar inside a proton, in which three quarks, each with a distinct charge, combine to have no net nuclear charge. The use of the word "color" reflects the similarities in the combination properties of color and nuclear charge but has nothing to do with the familiar chromatic properties of light.

don't have any appreciable effect on each other. Technically, we call this inability of quarks to leave the proton or neutron that contains them under normal conditions being "confined."

It turns out that the analogy between the strong force and a spring can be extended further. If you pull a spring or a rubber band hard enough, it will break. The strong force acts similarly. If you pull two quarks apart, the strong force resists more and more. But if you pull hard enough, the strong force "spring" will break. The distance at which the strong force spring breaks is about the size of the proton, which explains why the proton is the size it is. When the spring breaks, the quarks are then no longer connected and can move apart. Because of details beyond the scope of this book, these quarks are not "bare" quarks and cannot be seen like an electron that is knocked out of an atom. The idea is discussed in a little more detail in the text surrounding figure 6.8. Briefly, in the "breaking" of the strong force spring, this energy creates more quarks and antimatter quarks. (This is a consequence of Einstein's oft-quoted but rarely understood equation: $E = mc^2$.) In the end, quarks always travel in triplets or paired with a companion antimatter quark, safely ensconced in particles like protons.

The property of quarks most frequently mentioned is their mass. The mass of the various quark types spans a large range, from the up and down quarks with a mass about 0.004 that of the proton, to the superheavy top quark, with a mass of 173 times that of a proton. We have only a hazy idea as to what gives the quarks their respective masses, although it appears that one of the LHC's biggest triumphs to date has been to test the most popular mechanism that has been proposed to explain where they get their mass. This story is told in some detail in chapter 6.

One thing that is quite striking about quarks is the fact that there seems to be a recurring pattern in their appearance. For instance, the up, charm, and top quarks all have the same electrical charge, as do the down, strange, and bottom. Further, the up and down quarks are natural partners, in that they are the only quarks present in the stable proton and neutron. For this reason, as well as others, it is natural to group the quarks into three distinct pairs. We call these pairs *generations* and give each generation a number. The up and down quarks are generation I, while charm and strange quarks form generation II, and top and bottom quarks form generation III. The reason for three similar groups of quarks is quite mysterious and is probably telling us something profound, if we only had the wits to understand it. Perhaps the LHC will eventually teach us why this recurring pattern is present, and we will explore the question again in chapter 7. Table 2.1 summarizes what we know of quarks.

Leptons

While we have identified protons, neutrons, and electrons as components of atoms and quarks as components of protons and neutrons, we've not discussed the role of quarks in the electron. That's because there are no quarks in electrons. In fact, like the quark, the electron is thought to be fundamental, which is to say that the electron contains no smaller particles within it. Electrons have electrical charge like quarks do, but they do not have color charge. Because of this, each electron is not confined in the manner of quarks, which explains why they are not stuck in the nucleus but rather are free to orbit in the outskirts of the atom.

We said earlier that the universe in which we live can be built up by a proper mixture of up and down quarks and electrons. But we also know that there are two additional "carbon copies" (i.e., generations) of these quarks (e.g., charm and strange and top and bottom). Are there counterparts to the electron that might accompany these heavier quarks? Indeed there are. We have discovered two additional particles, called the *muon* and *tau*, that have the same electrical charge and general characteristics as the electron but are heavier. Like the word "candy," which we use generically when we don't need to specify exactly what sugary food we're talking about, there is a word that allows us to refer to all electrons and electron counterparts. This word is *lepton* and specifically a charged lepton to remind us that these particles carry electrical charge. The word lepton stems from the Greek *leptos* (for light). Like much of physics, Greek letters are used to symbolize these objects. The symbol for the muon is μ (the Greek letter *mu*), while the symbol for the tau is τ (for the Greek letter *tau*). Figure 2.6 and table 2.1 show how these charged leptons fit in to the overall particle picture.

While the electron (an electrically charged lepton) is a familiar particle, there also is a class of leptons that isn't so familiar. In the early 1900s, the study

Table 2.1. Names and characteristics of various subatomic particles

	Matter Particles: Quarks					
Generation	**I**		**II**		**III**	
Name	Up	Down	Charm	Strange	Top	Bottom
Symbol	u	d	c	s	t	b
Charge[a]	$+\frac{2}{3}$	$-\frac{1}{3}$	$+\frac{2}{3}$	$-\frac{1}{3}$	$+\frac{2}{3}$	$-\frac{1}{3}$
Mass[b]	~0.003	~0.005	1.5	~0.1	173	4.5
Discovered[c]	1964	1964	1974	1964	1995[d]	1977
Lifetime[e]	∞	∞	10^{-12}	10^{-8}	10^{-24}	10^{-12}
	Matter Particles: Leptons					
Generation	**I**		**II**		**III**	
Name	Electron	Electron neutrino	Muon	Muon neutrino	Tau	Tau neutrino
Symbol	e	ν_e	μ	ν_μ	τ	ν_τ
Charge[a]	−1	0	−1	0	−1	0
Mass[b]	~0.0005	~0	0.1	~0	1.8	~0
Discovered	1897	1956	1937	1962	1975	2000
Lifetime[e]	∞	∞	10^{-6}	∞	10^{-13}	∞
	Force Causing Particles					
Force	**Strong**	**Electromagnetic**	**Weak**			**Gravity**
Name	Gluon	Photon	Z zero	W plus	W minus	Graviton
Symbol	g	γ	Z^0	W^+	W^-	G
Charge[a]	0	0	0	+1	−1	0
Mass[b]	0	0	91	80	80	0
Range[f]	10^{-15}	infinite	10^{-18}			infinite
Strength[g]	1	0.01	0.00001			10^{-40}
Color	Yes	No	No			No
Discovered	1979	1905	1983			No
Particles affected	quarks	quarks, charged leptons	quarks, charged leptons, neutrinos			all

[a]Electrical charge relative to a proton, which has a charge of +1.

[b]Mass in units, so the mass of a proton is 0.94.

[c]The up, down, and strange quarks had been observed (but not recognized) before 1964, which was the year the quark hypothesis was proposed.

[d]The author was one of the co-discoverers of the top quark.

[e]The lifetimes listed here are in units of seconds and the lifetimes listed for the quarks should be taken as representative only, as a quark's lifetime depends on its environment.

[f]The range is listed in units of meters.

[g]The strength of all forces is referenced to the strong force.

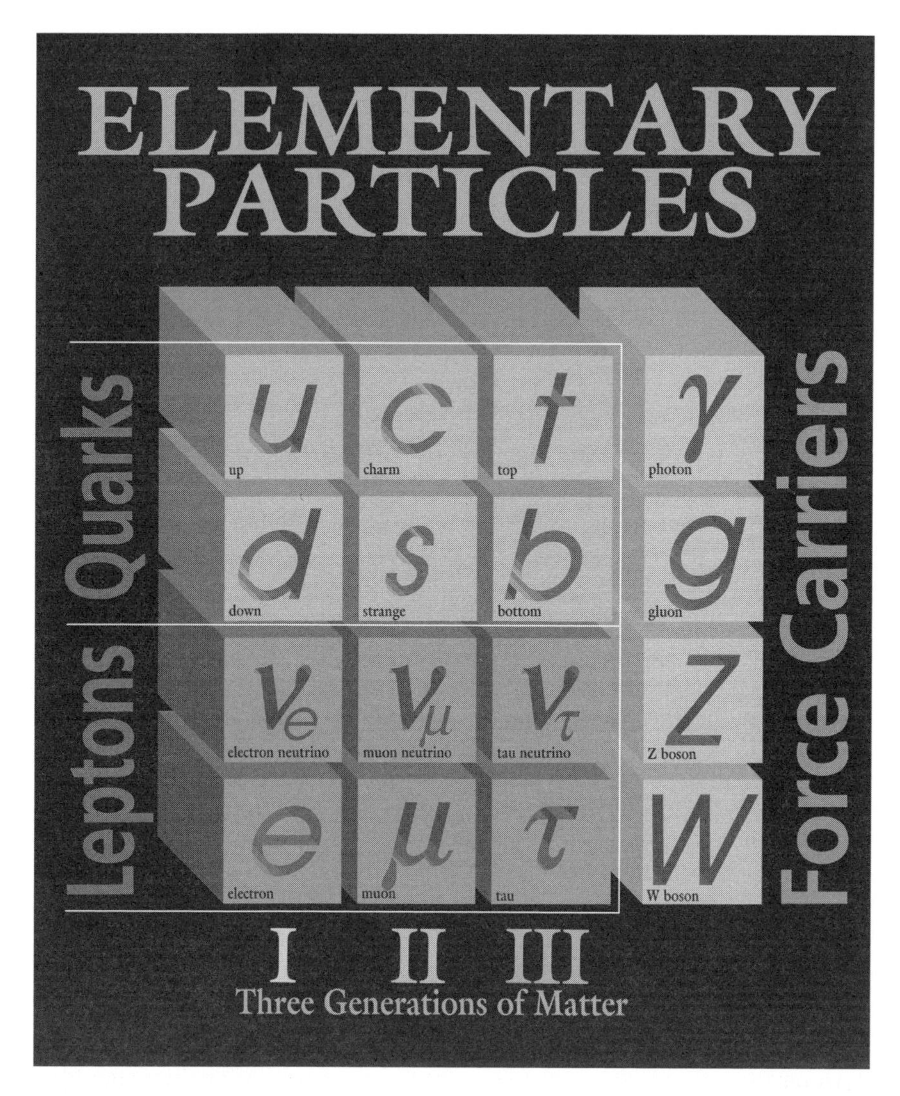

Figure 2.6. This table replaces the chemical periodic table as a list of the smallest-known building blocks and shows the particles that are responsible for the four types of forces. Figure courtesy of Fermilab.

of radioactivity was all the rage. *Radioactivity* is the decay, or transmutation, of the nucleus of the atom of one element into the nucleus of another element. In those early days, a class of radioactivity (called beta decay) perplexed physicists. When physicists looked at the energy involved in beta decay, they found that the energy after the decay seemed to be lower than it was before. This fact flabbergasted physicists, as it was a fundamental tenet of physics at the time (and still is) that energy cannot be created or destroyed. Clearly something was awry.

The conundrum was solved in 1930 when Wolfgang Pauli realized that the radioactivity mystery could be explained if a particle was emitted in the process of radioactive decay that was both very light and electrically neutral. A name was proposed for the particle, *neutrino*, from the Italian for "little neutral one." (Actually the name came from the Italian scientist Enrico Fermi, not the Austrian Pauli. Pauli's term "neutron" came to mean something else.) The neutrino was a new class of lepton, called a "neutral lepton." It had many interesting properties, including a near-zero (but not exactly zero) mass, and interacted with matter very, very weakly. The neutrino was first experimentally observed in 1956. The symbol for a neutrino is ν, the Greek letter *nu*.

When Pauli proposed the neutrino, he didn't fully appreciate just how peculiar a particle it was. The reason the energy budget didn't add up in these peculiar radioactivity experiments was because the neutrino was carrying away some of the energy. Later experiments showed that neutrinos can pass through lots of matter without being detected. While the penetrating power of neutrinos depends somewhat on their energy, neutrinos of the energies typically seen in radioactive decay could pass through five light-years of solid lead with just a 50% probability of being detected. Five light-years is 30 trillion miles. So it's not at all surprising that the physicists doing those early radioactive decay experiments were unable to see the neutrino and were therefore confused.

While Pauli spoke only of a single kind of neutrino, in 1962 an experiment showed that there is more than one kind of neutrino, with an electron-type and a muon-type clearly identified. Naturally, physicists wondered if there was a tau-type as well, a hypothesis confirmed in 2000. In order to distinguish the three types of neutrinos, we write them with a subscript. See figure 2.6 for examples.

With the realization that there are three distinct types of neutrino, each with an affinity for a particular charged lepton, our catalog of the known types of matter particles is complete. Ordinary matter is made exclusively of up and down quarks, plus electrons and electron-type neutrinos. Why there should be two carbon copies (charm, strange, muon, and muon-type neutrino and top, bottom, tau, and tau-type neutrino) is not understood, but these twelve particles (six quarks, three charged leptons, and three neutral leptons) is the entire list of matter particles that we've discovered thus far.

Forces

While we've listed the particles of which we're aware, we've entirely neglected a crucial part of the story. After all, *something* keeps the planets circling the sun, electrons surrounding the nucleus of atoms, and the proton and neutron firmly

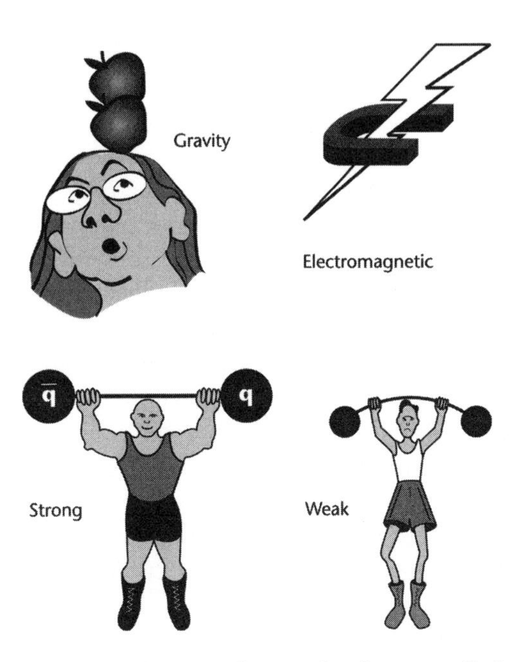

Figure 2.7. While four different forces are known, they have very distinct properties. Not only do they vary in strength, they also vary in their range and the manner in which they affect nearby matter particles. Figure courtesy of the Particle Data Group, Lawrence Berkeley Laboratory.

ensconced in their safe nuclear cocoon. These phenomena are governed by an idea called a *force*.

Forces can be simply defined as something that governs the motion of a particle. The force can attract or repel. Forces can even govern phenomena like radioactivity, which is kind of weird given the normal meaning of the word as, well, forcing something to move. In fact, we should use the word "interaction" instead of force to cover the radioactivity case. But the word force is so ingrained that we'll stick with it. Physicists currently know of four forces (figure 2.7). The most familiar of the forces is *gravity*, which keeps us solidly planted on Earth and governs the motions of the heavens. Ironically, this familiar phenomenon has most jealously guarded its secrets and remains the most mysterious force in the subatomic realm.

The second most familiar force is *electromagnetism*, which explains electricity of course, but it also explains magnetism, light, and all of chemistry. The electromagnetic force is much, much stronger than gravity and can cause both attraction and repulsion between two objects, while gravity is only attractive.

The other two forces are much less familiar. The *strong force* is responsible for holding the nucleus of the atom together, while the *weak nuclear force* is responsible for some kinds of radioactivity. As their names suggest, they have wildly different strengths.

Two important properties that distinguish the various forces are their range and relative strengths. Both gravity and electromagnetism have infinite range. In principle, every atom feels the effects of gravity from every other atom in the universe. In contrast, the strong and weak forces are only relevant over a very small distance and become essentially zero when the distances under consideration become larger than a proton.

With such different behaviors, no single number characterizes the forces' relative strengths. After all, two quarks separated by a distance just larger than the nucleus of an atom would feel no effect from the strong or weak force but would feel the effects of both gravity and electromagnetism. But once we get two particles close enough that all four forces come into play (say about the distance separating two protons in an atomic nucleus), we can compare their strengths. In doing so, we find that these strengths span an enormous range.

For instance, if we take the strong force to be the standard against which we compare the other three, we find the second strongest force, electromagnetism, is about a hundred times weaker. The third strongest force, the weak force, is about a hundred thousand times weaker than the strong force. The most familiar of the forces, gravity, is weak enough to be in a class of its own. Gravity is about 10^{-40} times smaller than the strong force. For those readers whose scientific notation is a bit rusty, remember that 10^{-2} is the same as 0.01. Thus 10^{-40} is a zero, followed by a decimal point, then thirty-nine zeros and a one. That's small! In fact, gravity is so weak that we've never been able to see any effect caused by gravity in modern particle physics experiments. Consequently, a quantum theory of gravity has eluded us. We simply don't know how gravity works in the realm of the ultrasmall. Further, the relative weakness of gravity is very troubling to physicists, and working out the reason for this weakness is something to which it is hoped the LHC might contribute.

While people have a feel for how gravity works, at the subatomic level, forces reveal a funny behavior. For "big" sizes, say about the size of a molecule, gravity is simply everywhere. Wherever you walk, gravity always pulls you downward, and there is no place where there is no gravity. In the quantum realm, things act differently. It turns out that, in the same way that atoms are small bits of elements, there are smallest bits of force. Each force has a characteristic particle associated with it.

The idea that a force like gravity could consist of small particles could be seen as counterintuitive, so let's talk about that idea a bit. Think, for a moment, about wind. Wind blows in your face, keeps a kite in the air, or pushes an empty can down the road. Wind exerts a force and is therefore analogous to other forces, like gravity or the electric force. We think of air as some kind of fluid perhaps, but in any event, it is something that is everywhere just like gravity.

However, we also know a fair amount about chemistry, atoms, and molecules. We know that air actually consists of molecules of oxygen, nitrogen, carbon dioxide, and the like. Thus the wind in your face is actually caused by uncounted molecules hitting you. Similarly, all of the forces at the subatomic level are treated as consisting of little particles of force.

Like much of particle physics, the names of the force-carrying particles are silly or simply mysterious. The particle causing the strong force is called the *gluon*, because it "glues" the nucleus together. The *photon*, familiar as light, is the particle carrying electromagnetism. Both the gluon and the photon have zero mass, but this isn't true for the weak force. Indeed, three types of particles cause the weak force: the electrically neutral Z^0 (simply called the Z boson) and two particles with electrical charge, W^+ and W^-, pronounced "W plus" and "W minus," with the "+" and "–" to show that they have the electrical charge of a proton (+) or an electron (–), respectively. These three particles are quite heavy, with each one carrying a mass nearly a hundred times heavier than a proton.

The fourth force, the quantumly mysterious gravity, is thought to be caused by a particle too. This particle is called the *graviton*. The graviton has never been observed, and you should regard with suspicion any claim to its properties. However if it *does* exist, we are able to work out what some of its properties must be. For instance it must be electrically neutral and have zero mass. Someday the graviton might be observed, and there's a Nobel Prize in it for the clever person who manages it. However, given gravity's weak nature, this prize is not likely to be claimed any time soon.

While table 2.1 lists the known quarks, leptons, and force-causing particles and brings us to the very frontier of knowledge, there is one little wrinkle that has not yet been mentioned. Even though we think the handful of particles and forces we've described thus far is all that's needed to describe our universe, it turns out that there is a duplicate for every particle listed. Our next frontier topic concerns a mirror image of our familiar matter. This mirror matter is called *antimatter* and is one of the phrases popular with science fiction buffs that is science and isn't fiction.

Antimatter

The simplest description of antimatter is that it is the opposite of matter. Take some antimatter, add an identical amount of matter, and they both disappear in a blinding release of energy. The amount of energy released is huge compared with the amount of matter and antimatter involved. To give you a sense of scale,

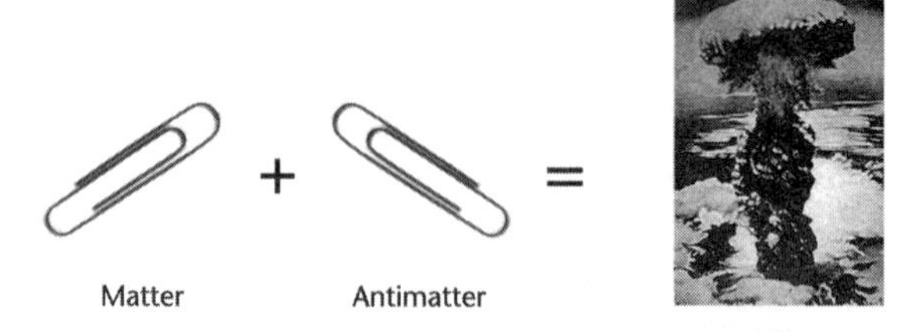

Figure 2.8. The energy content of matter and antimatter is surprisingly large. A matter paper clip weighing a gram, combined with an equivalent antimatter paper clip, would release energy comparable to the atomic bomb that devastated Hiroshima.

if you took a paper clip made of matter and let it touch a paper clip of antimatter, the energy release would be about the same as the 1945 atomic explosion at Hiroshima (figure 2.8).

Antimatter was predicted in 1928 by Paul Dirac. Does it really exist? The answer is a most emphatic "Yes!" The antimatter electron (called the *positron*) was discovered in 1932. The antiproton was observed in 1955, while the antimatter neutron waited until 1956. While protons and neutrons are made of quarks, their antimatter counterparts are made of antiquarks; for example the antiproton consists of two antimatter up quarks and one antimatter down quark. Antimatter particles have the opposite electrical charge of their matter counterpart; for instance, while the proton has an electrical charge of +1, the antimatter proton (the antiproton, $\bar{p}$) has an electrical charge of −1. It is even possible to combine an antiproton with an antielectron to make antihydrogen. In principle, all anti-elements could be made, although scientists have only been successful in making antihydrogen and antihelium. (A typographical aside: Throughout the book you will notice all sorts of letters with lines over them—$\bar{b}$, $\bar{p}$, $\bar{q}$. These refer to antimatter versions of a given particle.)

We have now observed antimatter counterparts for every type of quark and lepton. The simple existence of antimatter is interesting, but antimatter presents to us a truly fascinating mystery. To appreciate this mystery requires that you know two facts. First, you need to know that, when we make antimatter in our laboratories, it always comes with an identical amount of matter. Always. Make an antimatter up quark and you must simultaneously make an up quark. We never make an antimatter particle without a corresponding matter particle.

The second fact that one must consider is perhaps obvious but is extremely baffling. This fact is the simple observation that in everyday life, we just don't see antimatter anywhere. There's nothing in our understanding of antimatter that excludes antimatter stars, antimatter planets, or even antimatter people. As long as these things were kept isolated from matter, they should exist. And yet they don't. Nowhere in the universe, as deep as our telescopes can see, do we see any substantial chunk of antimatter.

So why is that? Nobody really knows. This doesn't mean that we know nothing about the subject but rather that the experiments done to date have not taught us the entire story. We expect the experiments of the LHC to shed light on the mystery, particularly the LHCb experiment (described in chapter 4).

With the introduction of the quarks, leptons, force-causing particles, and now antimatter, we have discussed everything we know. If one takes the particles from the first generation and tosses in the force-causing particles, we can build everything we see in the universe: from galaxies to ice cream. Toss in the particles from generations II and III and we can explain the results of all experiments ever conducted in our huge particle accelerators too. We call the theory that includes all these ideas the standard model of particle physics. A final component of the standard model is the Higgs boson, and it was an unproven hypothesis when the LHC began operations. Indeed discovering the Higgs boson or falsifying the idea entirely was a primary goal of the LHC. Chapter 6 describes in detail the outcome of those efforts.

With such a broad set of phenomena that we understand as well as we do, scientists are justifiably proud. To be able to take a handful of different types of particles and paint the tapestry of the cosmos is not a small feat. However, one should not be left with the impression that such an accomplishment has not left profound mysteries. In fact, for all our achievements, there's still a lot to do. Having focused our efforts on describing what we know, let's now shift our attention to the accelerator itself and the detectors arrayed around it and what we hoped to find out—and actually did find out—from them. After that, we'll be ready to launch into the unknown and learn how scientists are tackling some of the grandest puzzles of physics.

Before we move ahead, we must make a concession to the realities of having a diverse readership. This concession is to admit that some readers are more fascinated by scientific equipment than others are. Since this book is about the Large Hadron Collider, it would be inexcusable to not talk about the accelerator itself and the detectors arrayed around it. The inner workings of these machines are described in chapters 3 and 4. However, some readers just aren't gadget people. These are the same people who want their cars to run and their cell phones to work without a shred of curiosity as to the innards of these devices. If you're such a person, you can look for the return of our narrative of discovery in chapter 5. But, even if you're not a fan of blinking lights and only skim the next two chapters, be sure to look at the war stories at the end of chapter 4. For those who love knowing how things work as much as I do, let's take a look at some of the coolest scientific equipment ever built.

3

Accelerators and the LHC

In the previous chapter, we spoke of what we know about the world, and in chapters 5 through 8 we will talk of the LHC's successes and future prospects. Without an understanding of the equipment involved, it isn't possible to appreciate the magnitude of the effort going into the research. This chapter concentrates on the particle accelerator itself—the Large Hadron Collider—as an instrument of discovery. The LHC is an extremely complicated device, 17 miles (27 kilometers) in circumference and comprising 1,232 primary magnets that take 4,300 miles (6,920 kilometers) of wire to make. That's enough wire to stretch from New York City to Las Vegas and back. But before we focus on the specifics of the LHC, we need to spend some time understanding the physical principles and technical ideas that go into the design of a modern particle accelerator. The first part of this chapter discusses the main ideas involved in the design of any modern accelerator, while the end of the chapter concentrates on the specific details of the LHC itself.

Fundamentally, the problem is this: we want to procure a source of protons and accelerate them to the outrageous velocity of 670 million miles (a billion kilometers), per hour which is about 99.999999% the speed of light. Then we need to make a beam of these protons no wider than the width of a human hair and guide the beam in a circular path for a day or so, during which a proton will travel 16 billion miles (26 billion kilometers). In order to collide the beams, we need to steer them with exquisite precision. To give a sense of scale, the precision needed is equivalent of taking two sewing needles, separating them by 6 miles (10 kilometers), shooting them at one another, and having them collide at the halfway point.

All of these requirements sound pretty daunting, but the situation is made worse by the need to have two such beams and to require them to hit exactly head-on at specific times and places in the accelerator (figure 3.1). Oh, and by the way, this needs to be fairly simple to do, relatively quick, reliable, and at a manageable cost. I don't know about you, but the whole thing seems to be hard enough to verge on the impossible. And yet the CERN laboratory has

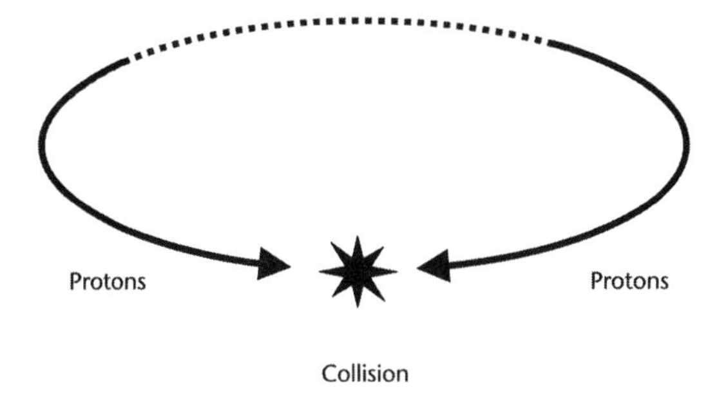

Figure 3.1. The highest energy accelerators rely on a combination of electric and magnetic fields to guide charged particles in a circular path and with increasing velocity. In the LHC, protons are collided at velocities approaching the speed of light, causing the highest temperatures ever achieved by humans and allowing scientists to re-create the conditions of the early universe and to study the building blocks of matter in considerable detail.

accomplished just this. If you're not impressed by this, you've simply not understood the magnitude of this Herculean task. The Augean stables were simple by comparison.

Acceleration

So let's start with the basics. If you have a bunch of protons, how do you cause them to go fast? A slingshot? Draft a pitcher from Major League Baseball? Attach them to a three-year-old and feed him sugared breakfast cereal? Well, while all of these approaches might have their merits, the reality is a bit more practical.

If you have an object at rest and you'd like it to move, you need to use a force. We discussed the four forces in the previous chapter. They are listed here ordered in strength from highest to lowest: strong, electromagnetic, weak, and gravity. Each of these forces affects different properties of matter. The strong force interacts with particles carrying the color charge, such as the quarks that make up the protons. However, since the protons themselves have no net color, that rules out using the strong force. The weak force is, well, weak and only works over a short range, and so that's out too. The electromagnetic force interacts with particles carrying electric charge. Since the proton has electric charge, the electromagnetic force is a candidate. Gravity affects particles with mass, and protons are massive particles. Further, gravity is a long-ranged force and is therefore a candidate. However, as we noted in the first chapter, gravity is extremely weak, which means it is rather unsuitable for an accelerating force. This leaves electromagnetism and specifically electric fields to provide the impetus that causes the proton to move.

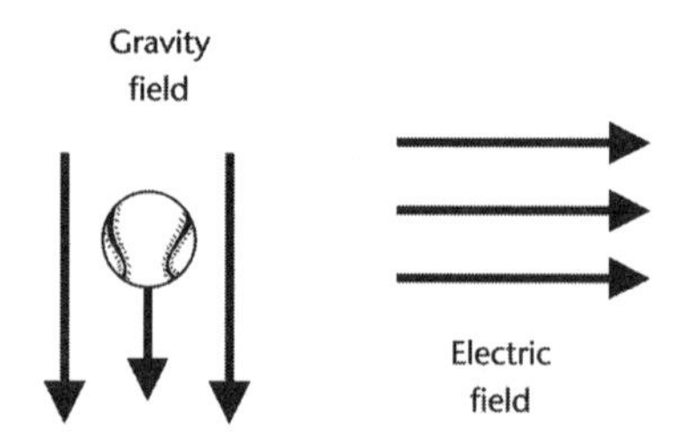

Figure 3.2. Electric fields and gravity fields can both cause objects to move. Electric fields are unlike gravity, in which the relevant feature of an object, such as a baseball, is its mass. For electric fields, it is the electric charge of the object that interacts with the field. The direction of electric fields can be manipulated at will, allowing charged objects to be accelerated in arbitrary directions.

It's actually pretty easy to understand how an electric field causes a proton to move. Fundamentally, it's a lot like gravity, with which we have ample familiarity. By way of analogy, hold a baseball up high and let it go. Gravity acts on the mass of the ball and it drops. Technically, we would call gravity a "gravity field" to indicate that the effects due to gravity are apparent over a large area.

Similarly, we can make an electric field that will interact with the electric charge of the proton and force it to move. We know how to make an electric field point in any direction we want, and so naturally we orient the electric field so the proton is driven through our accelerator (figure 3.2).

So how does one make an electric field? There are lots of ways to do it, although not all are practical choices for a particle accelerator. But to give a basic idea of how it works, you can take a rubber balloon and rub it on your shirt. (A mylar balloon won't work.) Then take the balloon and run it just above your arm. You will find the electric field from the balloon tugs your arm hairs. This experiment works best on a cool and dry day.

However, the simplest way to create an electric field that is useful for accelerator purposes is to take two plates of metal and connect them with wires to a battery. Between the two plates, an electric field will be set up.

Now when one makes such an electric field with a simple 1.5 volt D-cell battery, the electric field isn't very strong and the proton isn't accelerated very much. Since we want to accelerate protons to extremely high speed, this is a technical problem that must be overcome. There are two solutions. The first is to simply use a stronger battery. This is the approach used in old-style TVs (i.e., cathode ray tubes, on which some of us first watched *Gilligan's Island* or *Star Trek*), which used a battery in excess of 10,000 volts.

Another approach is to take a bunch of particle accelerator units like those

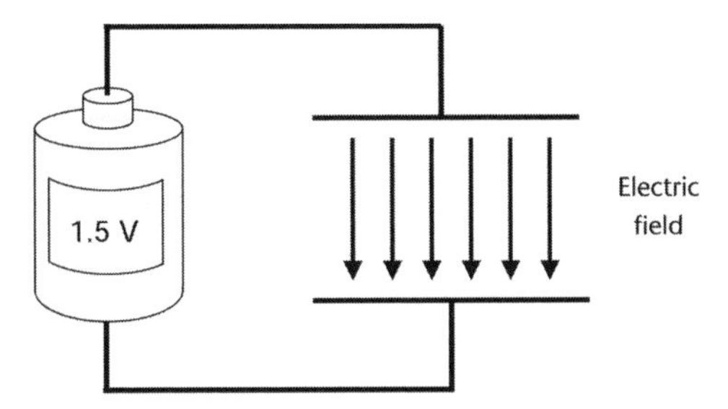

Figure 3.3. It is a simple matter to create an electric field. Taking a simple battery, two wires, and two parallel plates of metal, you can set up a uniform electric field, which is the first stage of creating a particle accelerator.

sketched in figure 3.3 and stack them up. The particle would then be accelerated a little bit by the first one, more by the second one, even more by the third one, and even more by subsequent units. If one wishes to make a gravitational analogy, if the accelerator apparatus illustrated in figure 3.3 is equivalent to dropping a marble from a height of one foot, then two such sets of equipment is equivalent to dropping it from a height of two feet, and so on.

Modern particle accelerators use both techniques. Current technology can make electric fields that are equivalent to attaching a 50 million volt battery to two metal plates separated by 3.2 feet (about one meter). (Note that electric fields of that strength are quite difficult to achieve.) However, remember that the point of this exercise is to use this technology to accelerate protons to a desired velocity so they will collide with adequate energy. Consequently to understand accelerators, we need to understand how voltage and energy are related. Particle accelerators use a very convenient unit of energy. Rather than the familiar unit of kilowatt-hours you see if you read your electric bill, or even joules from a science class you've taken, particle accelerators use "electron volts" (or eV, note both letters are pronounced, i.e., "ee-vee"). One electron volt is the energy an electron (or proton) gets when it is accelerated by a one-volt battery. Thus electrons in an old-style TV, with its 10,000-ish volt battery, are accelerated to an energy of 10,000 electron volts. The LHC, with its ultimate beam energy of 7×10^{12} eV—that is, 7,000,000,000,000 eV, or 7 TeV (trillion electron volts)—would require the equivalent of 7×10^{12} volts of batteries. This allows us to estimate how long a simple accelerator might be. For instance with the accelerating voltage mentioned above (50 million volts per meter), it would require 84 miles (135 kilometers) of batteries to achieve. Technically, you would accomplish this by taking 140,000 of these "two metal plates and a battery" contraptions and laying them end to end.

Good Vibes

So far, everything I've told you about how an electric field can be used to accelerate a particle is true, relevant, and I hope interesting. There's just a tiny little thing I forgot to mention. Electric fields in accelerators aren't made by the "two metal plates and a battery" idea. It's not that such an approach couldn't be made to work. In fact, early accelerators adapted this approach. The reason it's not used very often anymore is because there turns out to be an easier and less expensive way to make strong electric fields. To understand how requires a little intellectual detour to America's rural cultural heritage and to my Uncle Jeb's country jug band.

For those of you not lucky enough to have grown up in the deep country, a jug band is a traditional and time-honored way to make country style music. Such a band consists of various homemade instruments, including a washboard, a mouth organ, a contraption consisting of a metal washtub, broom handle, and twine and, most important for our purposes, a guy blowing across the top of a big moonshine jug. If he blows just right, the jug will emit a baritone sound that I've decided sounds like "huv." This brings us to the most important point in our detour. If the guy blows too fast or too slow, the jug doesn't make a loud sound. However at the perfect "blow speed," the jug emits a loud sound. The sound emitted by the jug is much, much louder than the sound of the guy blowing. You can reproduce this phenomenon yourself by blowing across the top of a large empty soda bottle.

So, getting back to particle accelerators, it turns out that one can make the equivalent of a jug for electric fields. If you make a hollow metal container of the proper shape (the technical term is a *cavity*) and "blow on it" with radio waves, you can make very strong electric fields. Just how this works is a bit complicated but is essentially the same phenomenon as the jug in Uncle Jeb's band. Because the way you "blow" on the cavity is using radio waves, we call this technique RF (for radio frequency).

Designing the precise shape of a cavity to make a strong electrical field is a complex and fascinating endeavor. Obviously, scientists want to make the strongest electric field in the smallest volume . . . to get the most bang for their buck, so to speak. Just like racecar mechanics will use every trick at their disposal to increase the top speed of their car by a single mile or kilometer per hour or to get a better cornering speed, accelerator designers exploit every technical trick in their book to make the most perfectly designed cavity to get the strongest electric field.

Luckily, we don't need to know all of these subtleties. Some cavities are shaped like a big bagel (figure 3.4). Radio waves are aimed into the side of the

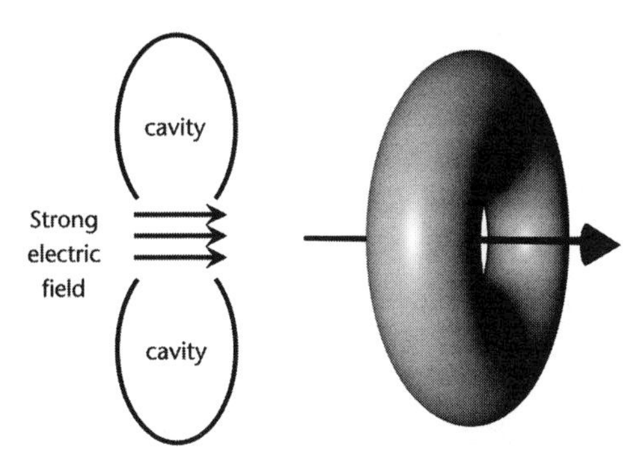

Figure 3.4. The electric fields used in modern accelerators are set up using a toroidal (or donut-shaped) structure. By injecting radio waves into the donut, strong electric fields are set up in the donut's "hole." ***Left*****, cross section of an accelerating structure;** ***right*** **three-dimensional version of the same.**

bagel, and the strong electric field (the loud spot in the jug analogy) is at the hole of the bagel. This works out beautifully, because we can aim our particles through the center of the bagel where the electric field is very strong.

This little detour into jug bands and making electric fields may seem a bit out of place, as this is a technical detail that one would ordinarily gloss over in a book of this type. However, this detour is essential in getting a good "gut feeling" about just what is going on in the LHC.

Sound is a vibration. If you get near a pipe organ (which is essentially a very powerful version of our modest jug), you can actually feel the vibration when a base note is played. Similarly, the radio waves that are used to excite the accelerator cavities are also vibrations (hence the "F" in RF). So this gets us back to the point of the detour. The electric fields in a particle accelerator are vibrating.

Note that this is somewhat different from the electric field and gravity analogy we drew a little while ago. After all, gravity is constant. But the electric field in a particle accelerator is constantly changing. It has a maximum, drops to zero, and even reverses direction entirely (although we'll ignore that particular fact in the following discussion, since it just unnecessarily complicates things). The upshot of all this is that we need to carefully time our beam's passage through the electric fields to occur when the field is strongest. That's the time when the charged particle will be accelerated the most. A consequence of this is that the charged particle will only pass through the electric fields at particular times.

We can most easily envision how this works by watching surfers. If you watch a surfer as she's starting to move quickly, you'll note that she's always in a particular spot on the wave. She's in front of the wave, about halfway up, somewhat closer to the top. That's the "sweet spot," where you get the best ride.

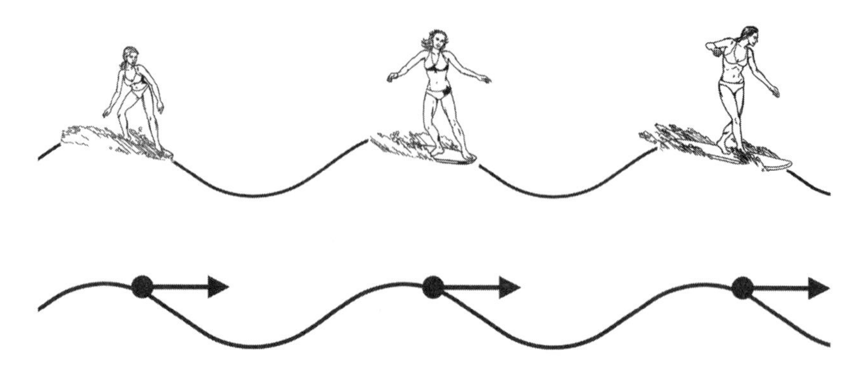

Figure 3.5. The minimum separation between adjacent surfers is the separation between adjacent waves. Similarly, particles in an accelerator are separated by the wavelength of the radio waves that generate the electric field. In order to accelerate efficiently, particles must be located on a specific spot on the wave. Surfers courtesy of Dan Claes.

If the surfer were on the back side of the wave, she wouldn't move, nor would she move if she were in a trough.

This observation has interesting consequences. Suppose you had a bunch of highly efficient surfers who wanted to take advantage of every wave. As each wave went by, each surfer would catch it. Thus while surfing, the surfers would not get closer than a wave-separation apart (figure 3.5). Similarly, when accelerating particles, the particles in one wave are always separated from ones in the adjacent wave. The separation depends on a particular accelerator's design, but in the LHC particles riding adjacent waves are separated by about 25 feet (7.8 meters). Note that particles don't *have* to ride each wave (for instance in the Fermilab accelerator, particles were only put in every twenty waves or so, and during 2010 to 2013 particles in the LHC occupied every other wave), it's just that they can't be closer than the separation of adjacent waves.

Returning from our detour, we can get back to the subject of accelerator length. A few pages earlier, we showed that the accelerator length of the LHC using the technologies discussed thus far could be over 84 miles (135 kilometers) long. Now making a particle accelerator that is that long is not without its challenges. The most pragmatic challenge is cost. Basically, if you double the length of your accelerator, you double the cost. So pretty much any trick you can play to shorten an accelerator's length is worth doing.

Round and Round We Go

In the 1940s, scientists had a good idea. What if you could somehow use the same accelerator over and over again? On the face of it, it seems as though we're talking about something akin to magic. Suppose you could drop a ball from a

Figure 3.6. A tetherball provides an accessible metaphor for the process of accelerating a charged particle in a synchrotron. In the game of tetherball, a person hits the ball (*left*) and then waits for the rope to guide the ball around in a circle for a subsequent hit (*right*). In a synchrotron, a short region with an electric field accelerates the charged particle, while magnetic fields guide the particle back for another acceleration episode. Figure courtesy of Dan Claes.

height of 10 feet (3 meters) and when the ball got to the bottom you could use magic to make it disappear from the bottom (*poof!*) and reappear at the top (but with the same velocity it had at the bottom). If you could repeat this ten times, then you could accelerate a ball with a 10-foot-tall tower to a velocity that would ordinarily take a 100-foot-tall (30 meter) tower without using magic.

Well, using magic in a physics laboratory is usually considered a poor choice, what with the rabbit fur and dove feathers getting into everything. No, the wizardry used to solve this particular problem was of a technical and not sleight-of-hand variety. This engineering magic is called a *synchrotron* (although there were predecessor technologies developed in the 1930s that used some of the same tricks). The principle that governs a synchrotron is essentially identical to the one that governs the game of tetherball (figure 3.6).

A tetherball is a ball attached to a rope. The other end of the rope is attached to the top of a tall pole, anchored deep into the ground. A person hits the ball, and the rope makes the ball travel in a circular path. The ball comes full circle, and the person can hit it again. The ball goes faster and makes another circuit. If the rope is connected to the top of the pole in such a way that it doesn't wrap around the pole, in principle you can get the ball going very fast by synchronizing (and hence the name) the orbit of the ball and the person hitting it.

In a particle synchrotron, the electric field "hits" the proton and accelerates it. However, the counterpart of the rope in the tetherball analogy is not provided

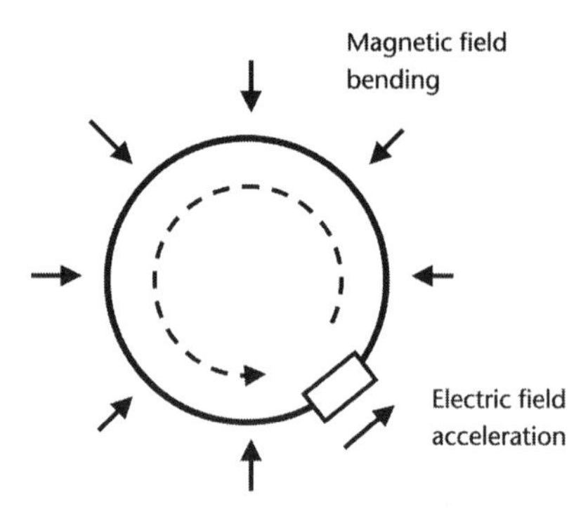

Figure 3.7. Schematic of a synchrotron. An electric field (the "hit" in the tetherball analogy) accelerates the charged particle and magnets (the rope) guide the particle in a circular path (denoted by the solid line) back to the electric field, where additional acceleration occurs. The arrow in the electric field region indicates acceleration in the direction of motion, while the inward-pointing arrows show the magnetic field guiding the particle in a circular path. The dashed line shows the sense of the particle's motion. In the case of the LHC, there are essentially two accelerators, each making a beam circulate in opposite directions.

by electric fields but rather magnetic ones. Electric fields and magnetic ones are two facets of the underlying phenomenon called *electromagnetism*. While we noted earlier in chapter 2 that electricity and magnetism are really the same thing, in an engineering context we treat them differently.

Magnetic fields are handy in synchrotrons, because charged particles like protons travel in a circular path when moving in a magnetic field. Figure 3.7 is a simple schematic of a modern particle synchrotron. Particles are accelerated by an electric field over a short distance and are then guided by magnetic fields in a circular path back to the electric field region for another round of acceleration.

Electromagnets

When I say that magnets are used to make particles orbit inside our accelerators, people often have different mental pictures of the magnets involved. Some think about the equivalent of the magnets that keep children's art on the front of the refrigerator. Others think of the horseshoe magnets that they played with in science class in elementary school. While both of these techniques create magnetic fields, particle accelerators use a more industrial approach: they use electricity to create an electromagnet to make the magnetic field.

At its simplest, one makes an electromagnet by wrapping wire many times around a chunk of iron. Then you connect the wire to a battery (figure 3.8). The electric current flows through the wire and makes a magnetic field. The iron isn't strictly necessary, but it acts as an "amplifier," which greatly increases

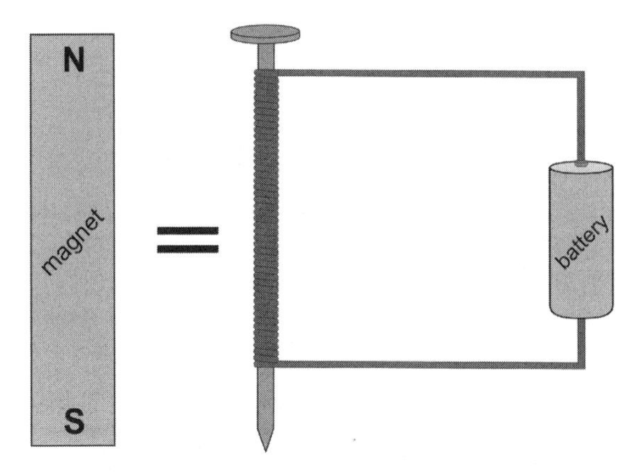

Figure 3.8. A common lesson in elementary school science is how to make an electromagnet. A common nail, wrapped in wire attached to a battery, becomes a magnet that can be turned on and off at will. The existence of the nail is not crucial, although the iron in the nail helps make a stronger magnet. Figure courtesy of Barry Panas.

the strength of the magnet. You can easily make an electromagnet at home by taking a handful of iron nails and taping them together into a bundle. Then wrap the bundle of nails with wire. The more loops the better, but twenty or so should do. Connect the ends of the wire to a battery, and you should be able to pick up bits of metal like nails and paper clips. A scaled-up version of this simple demonstration is what is used in junkyards to pick up beat-up cars and move them around.

The magnets in particle accelerators are quite similar in principle, although the engineering is a bit more precise. However, a coil of wire, a chunk of iron, and a battery are present in most accelerator electromagnets.

Why use the electromagnet technique? The first reason is magnets of this type can be very strong. Second, since you can shape the iron and wrap the wires in any way you choose, you can have great control over your magnet, choosing where the magnetic force must be strong and where it can be weak. The third and possibly best reason to use electromagnets is the presence of the battery. By changing the strength of the battery, you can alter the electric current in the wire, which in turn can vary the strength of the magnet. You can turn the magnet off, run it at full strength, or anywhere in between.

This brings us to an important point. The ultimate source of the magnetic field is the electric current passing through the coil of wire. More current means a stronger magnetic field. It's that simple. However, in general, most materials do not let current pass through them unimpeded. Materials resist the flow of current, and different materials resist the flow of current differently. Even copper, used in the wires in your house precisely because of its low resistance to the flow of electric current, does not let current freely flow through it.

While copper is a good material to be used to make electromagnets, it would be even better if new materials could be found that would resist the flow of electrical current even less. It turns out that there is a way to reduce a material's resistance to electrical current and to eventually eliminate it altogether.

It Doesn't Get Much Cooler Than This

In 1911, Heike Kammerlingh Onnes was studying the electrical properties of materials as they were cooled. It had long been known that cooling a material reduced that material's resistance to the flow of electrical current. However, Onnes was a refrigerator specialist and nobody could cool like he could. He was the first person to cool helium enough to cause it to turn into a liquid. This is the same helium that fills balloons. Helium is the coldest liquid ever discovered and turns from a gas into a liquid at –452°F (–269°C)—pretty close to absolute zero.

For his electrical experiments, Onnes was cooling mercury and watching its electrical behavior, which was acting normally. Specifically, the electrical resistance of mercury was predictably dropping as the temperature was lowered. However, when the mercury's temperature got to –452°F (–269°C), an unexpected thing occurred. Suddenly it no longer resisted the flow of electrical current at all. A new phenomenon called *superconductivity* had been discovered. Rather quickly, many other materials were shown to be superconducting at similarly low temperatures. The fact that the temperature at which helium liquefied and mercury became superconducting is the same has no particular significance. Other materials become superconducting at different temperatures, for example lead at –447°F (–266°C) and tin at –453°F (–269°C).

Superconductivity was an interesting effect for decades (and explained in 1957) but became relevant to particle accelerators when physicists needed to make the strongest magnets possible. For an electromagnet to be strong means that a lot of electrical current must flow in it. And, of course, providing lots of electrical current is easiest when the electrical resistance is lowest. So naturally they decided to see if they could make large electromagnets on an industrial scale from wires made of superconducting material.

The technical challenges were formidable, but eventually it was solved. In 1983, the Fermilab Tevatron, the first large accelerator made with magnets using superconducting technology, began to be commissioned.

Let's consider for a moment just how impressive a feat it is to have designed and built such a magnet. To do so, we need to understand a type of magnet called a *dipole*. Bending magnets are called dipoles for reasons that we look at

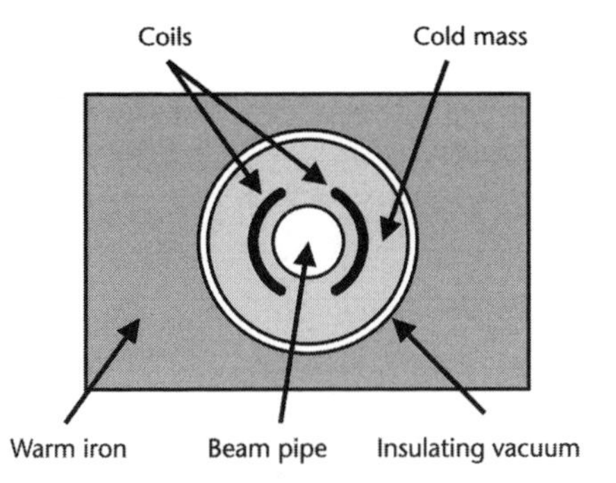

Figure 3.9. The view from the end of a typical dipole magnet. Wires are wound in a specific pattern, resulting in a vertical magnetic field. This causes a beam of particles travelling perpendicular to the page to be deflected left or right. This particular magnet is a superconducting one, which requires the coils of wire to be cold. The temperature specifications are the reason that the inner and outer portions of the magnet are separated by a vacuum. Just like an ordinary Thermos bottle, a vacuum is a poor conductor of heat. Line drawing courtesy of Barry Panas.

later in the chapter when we explore the types of magnets that do things other than bending the beam. Because this book is about the LHC, we would ordinarily use an LHC magnet as an example. However, the LHC magnets are more complicated than most. We'll show a simpler one at first. Figure 3.9 shows a two-dimensional cross section of a dipole that displays most of the essential features. The center of the dipole is a pipe through which the beam passes. The air in this pipe is pumped out, so the beam will not hit air molecules. The pipe is surrounded by tubes through which liquid helium can pass. On both sides of the hole are the coils of wire that make the magnetic field in the beam pipe. This is followed by strong structural material that keeps the coils from moving and then a series of cooling pipes, carrying ever-warmer material as one moves outward from the center of the magnet. The bulk of the magnet consists of an iron support that gives structure to the whole magnet and holds it together.

Figure 3.10 shows a cross section of the most common LHC magnet. The most striking difference from the ordinary magnet in figure 3.9 is that the LHC magnet has a passing resemblance to twins. There are *two* magnets buried in the surrounding metal and cooling equipment. This LHC bending magnet is about 55 feet (17 meters) long and weighs 35 tons (32,000 kilograms). The accelerator needs 1,232 of them. The coils of wire at the center of the magnet take miles of wire to make and have to be cooled to about –456°F (–271°C), but the outer surface of the magnet needs to be room temperature. That means the center of the magnets needs to be surrounded by what is effectively a high-tech Thermos

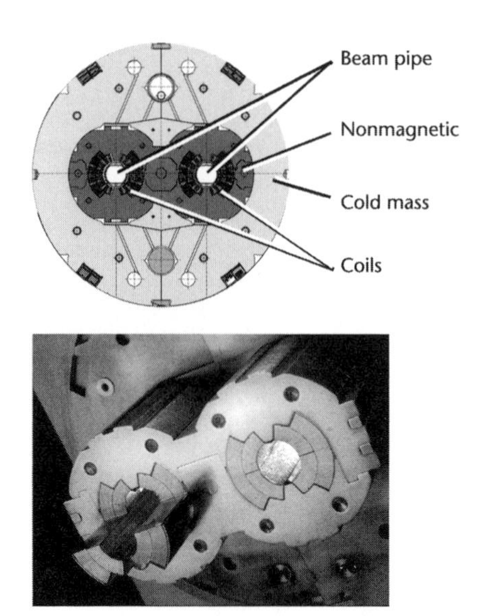

Figure 3.10. Real LHC dipoles are more complex than the magnet depicted in figure 3.9. Each magnetic structure actually contains two dipoles and not just one. *Top,* a basic schematic; *bottom,* a photo of actual LHC coils. Figure courtesy of CERN.

bottle. Even just sketching the basics of such a magnet reveals a daunting technological challenge, but of course a real-world magnet is even trickier.

Particle Beams

Up until now, we've discussed the challenging task of accelerating particles to near the speed of light and causing them to orbit in a circular path. However, let's not lose sight of why we have undertaken such a challenge. We want to collide a beam of subatomic particles head on with another such beam. Let's spend some time talking about what a beam of particles really is.

To begin with, a beam of particles requires more than one particle. The numbers vary depending on the accelerator, but having ten trillion protons in your particle accelerator at one time is not at all unusual. And, of course, when you have a situation with more than one object, things gets complicated very quickly. By way of example, imagine driving the freeways of Los Angeles when you're the only car on the road. Now think about the traffic on a Friday afternoon at rush hour, and you can easily see that the situation is far messier when you have more than one car (or particle!) involved.

Soon, we'll consider how the simple electric and magnetic fields mentioned earlier in the chapter must be modified to account for the real-world multiparticle beam. However, let's first spend a minute discussing its basic geometry.

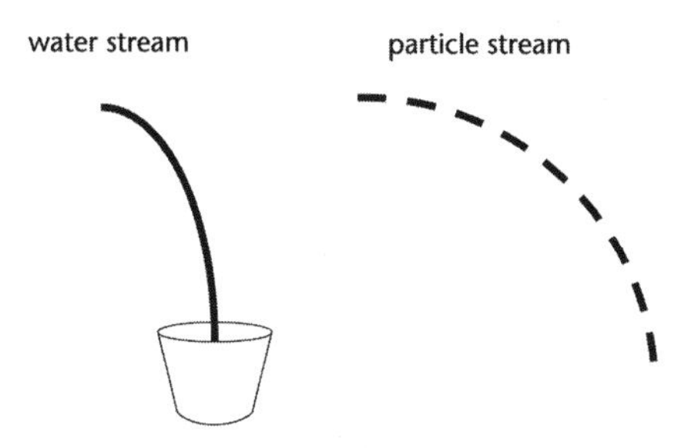

Figure 3.11. A stream of water (*left*) is continuous as it flows from hose into a bucket. However, a particle beam (*right*) is discontinuous, with clumps of particles that are separated from each other. This discontinuous nature is caused by the manner in which particles are accelerated.

Most people, when asked to describe the beam in an accelerator such as the LHC will say "Huh?" and give you a blank look. But even among people who have pondered such things, they'll usually describe something like a stream of water coming out of a high pressure water hose. This picture is right in many ways, in that the beam starts at a point and travels in a straight line essentially forever. To imagine this, think of a laser pointer. If you point it at your hand, you see a dot of light about a quarter of an inch wide (6 millimeters). Now direct the beam at a wall 100 feet (30 meters) away and you see a spot that is still about the same size.

However, particle beams are not continuous. To see what we mean by continuous, imagine the water hose analogy mentioned above. If you could somehow instantaneously freeze the water stream, you would see that there was a continuous slug of water from beginning to end.

The particle beam is different because there are "gaps" in the stream. If you could "freeze" a particle beam, you'd see a clump of particles, then a big gap, then another clump, and so on. The reason the beam has this "feast or famine" structure has to do with technical details of the accelerating electric field, which was mentioned earlier in our discussion of surfers, in which the sweet spot was near the top of the wave.

The basic structure of a particle beam is the following. You have many bunches of particles. Each bunch is about a foot (30 centimeters) long and about a thousandth of an inch (25 micrometers) in diameter. Each bunch is separated from its neighbors by tens or hundreds of feet (say, three to many tens of meters). Further, each bunch can contain hundreds of billions protons. The particles don't create a steady stream, like water coming from a hose; instead they form a dotted line (figure 3.11). The specific numbers vary for any particular accelerator. The LHC is designed to have 2,808 distinct bunches, initially with 100 billion protons per bunch.

When discussing any accelerator, there are three important parameters. The first is the type of particles being accelerated. The second is the energy of the beams. The third (and next topic of discussion) is the number of collisions per second. More collisions per second are good.

In an accelerator like the LHC, two beams of particles are aimed at one another. In many respects, it's like two rifles shooting bullets at one another, with the hope that the bullets collide head on. In the case of the LHC, two bunches, each containing hundreds of billions of protons, are made to pass through one another, with the hope that maybe two protons will collide with sufficient violence to do something interesting.

Focusing on the Problem

So let's think about what is necessary to increase the probability that two protons will get close enough together to collide violently. The trick is to make the beam width as tiny as possible. To see what I mean, let's instead talk about two swarms of bees flying through one another. If you had two such swarms and wanted to make more likely a head-on collision between two bees, what would you do? You'd force the bees within each swarm to fly closer to their neighbors.

In figure 3.12, we see the effect of forcing the bees in a swarm to fly in tighter formation. Because there is less space between adjacent bees, as the swarms pass through one another, the probability of a head-on collision increases.

Of course, we are interested in a particle accelerator and not a bee accelerator, but the same concept holds. To increase the probability of "interesting" and violent collisions between protons, one can do two things. First, put as many protons in your accelerator as possible and, second, make your beams as narrow as possible.

The language of this aspect of particle collisions borrows heavily from light and lenses. We talk about the "luminosity" of a beam, which is analogous to the brightness in a beam of light. A beam of light can be made brighter by increasing the amount of light in your beam or by using a lens to focus it.

So this naturally leads us to ask the question "How does one make a particle lens?" or, more generally, "How does one focus a particle beam?" Answering this question requires that we return to our discussion of magnets. If you recall, the magnets that forced the particles to go in a circular path were electromagnets, consisting of coils of wire and chunks of iron. Each bending magnet consisted of one coil of wire. These "lens-like," or focusing, magnets use two coils. The one-coil (bending) ones are called *dipoles*, while the two-coil (focusing) ones are called *quadrupoles*. However, we will continue to simply call them bending and focusing magnets.

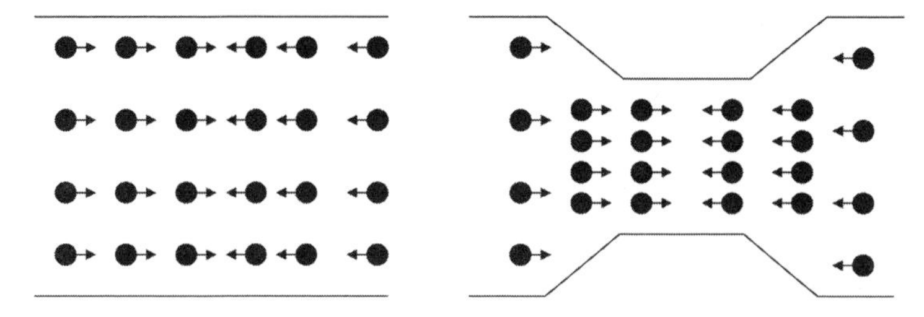

Figure 3.12. In a typical particle accelerator (*left*), "swarms" of particles pass through one another. In order to maximize the probability that two particles will collide, it is imperative to focus the beams (*right*).

While the superconducting bending magnets are the biggest and strongest types of magnet in any circular particle accelerator, the focusing magnets are no less crucial (indeed most accelerators also contain three-, four-, and five-coil magnets as well). The reason that the focusing magnets are critical is because lots of effects conspire to make the beam much wider than desired. The simplest of these are just the manufacturing and installation imperfections in the magnets that cause them to operate not as perfectly as the ideal case that has been discussed thus far. Another big effect comes from the self-destructive nature of the beam itself. Recall that the LHC beam consists of vast numbers of protons, which are positively charged particles. You may have heard "opposites attract." That may or may not work in dating, but it certainly works in electricity. Objects with opposite sign attract one another, while objects with the same-sign electric charge repel one another. Consequently, the electrical self-repulsion between the protons in the beam itself threatens to "blow the beam up" and widen it from its collision-optimized small size.

To give a sense of scale, we can mention the raw numbers of the different types of magnets in the LHC. There are 1,232 bending magnets, spread uniformly over the 17 mile (27 kilometer) circumference. In addition, there are 858 focusing magnets. There are also 7,210 smaller "correction" magnets, for a grand total of 9,300 magnets. These more-complex magnets are used to correct for the fact that not all protons in the accelerator have precisely the same energy, as well as other small effects. These three kinds of magnets work together like kindergarten teachers trying to get a bunch of kids moving as desired: "Go left! Do it again! Bunch up and stay together! Hey! Back in line! Yes, I'm talking to you!"

Thus far, we've discussed the LHC as if it were a single circular accelerator, and that's true in a strict sense. However, it's also misleading. It's more accurate to say that the CERN laboratory hosts an entire bevy of other accelerators, each crucial for the LHC's mission. The need for a series of accelerators is true for any modern laboratory. The logic is no different from what goes on in any automobile. Essentially, each accelerator can be treated as a different gear in a car. If you've ever driven a car with a manual transmission, you know that it is possible to take a stopped car and get it moving while only using the highest gear. It's just very hard to do, and you run the risk of frequent stalls. It's simply easier and more efficient to shift the car through a series of gears, each carefully tuned to match engine and wheel speed.

So too it is with accelerators. In the LHC, one has to accelerate stationary protons to nearly the speed of light: 186,000 miles (300,000 kilometers) per second. In principle, you could build a single accelerator that could span that entire range of energy, but the fact is that technical, engineering, time, and cost concerns make that a very poor choice. It is simply more efficient to have a series of accelerators, each tuned to a different energy. For example, the first might accelerate a proton from rest to 1% the speed of light and the second might go from 1% to 10%. The third might go from 10% to 80%, and so on. Such a "chain" of accelerators is what makes up the LHC complex, with the actual Large Hadron Collider being only the highest energy accelerator in the chain. However, without all of the steps in the chain, the LHC would be nothing more than an incredibly expensive tunnel connecting Switzerland and France.

Before we talk in detail about the LHC accelerator, we need to define one more pair of terms: fixed target versus collider mode. The first experiments were done in "fixed target mode," which means that they were all done by shooting a single beam into a stationary (i.e., fixed) target. This is akin to someone shooting at a wall or, given that we are discussing circular accelerators, akin to the sling with which David is reported to have slain Goliath. For those of you for whom biblical stories are a bit rusty, a sling is used to spin a rock in a circular motion. After the rock is moving at great speed, the rock is released and travels in a straight line toward its target. In the case of David and Goliath, the target was Goliath's forehead.

In fixed target experiments, a particle beam is accelerated using the methods we've discussed earlier in this chapter. The charged particles are then aimed at a target. The target can be anything, although it is typically a container of hydrogen, chilled until the hydrogen liquefies.

However, the LHC does not operate in fixed target mode. The "C" in LHC stands for "collider" after all. In such a configuration, two counter-rotating beams of charged particles are aimed at one another to collide head-on.

What are the advantages of colliding beams over fixed target experiments? The disadvantages are obvious. Aiming two beams at one another is hard, while in the fixed target case, you only have to aim your beam at a target, which you

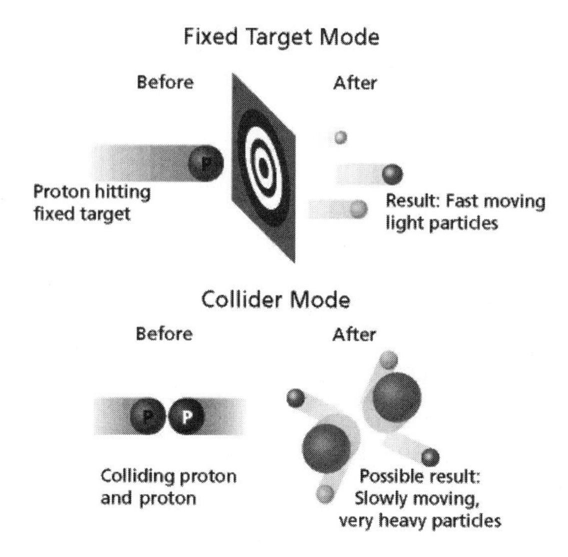

Figure 3.13. When a beam of particles is aimed at a stationary target (*top*), the result is collisions that are less violent than when two beams of particles are made to collide head-on (*bottom*). All modern high energy accelerators adopt the second approach. Figure courtesy of the Particle Data Group, Lawrence Berkeley Laboratory.

can make arbitrarily large. The phrase "hitting the broadside of a barn" comes to mind. So why a collider?

The reason is energy. When a projectile hits a stationary target, the debris from the collision ends up moving in the direction of the initial projectile. For a visual image, think of shooting at a watermelon or a car smashing into a tree farm full of saplings. However, in the collider configuration, the two projectiles can simply stop, like two cars involved in a head-on collision. When this happens, the result is that all of the energy of the collision can go into doing something "interesting." In our analogy, the interesting thing would be to thoroughly wreck the cars. In our particle physics world, interesting simply means revealing some new and rare physical phenomenon.

The difference in the available energy is surprisingly large. The LHC has two beams, each carrying as much as 7 trillion electron volts of energy. When particles in the opposing beams hit head-on, there are 14 trillion electron volts of energy available to possibly discover something new. However, if just one of these beams were to hit a stationary proton, the useful energy would not be the full 7 trillion electron volts of energy. Rather, the energy available would be a meager 0.1 trillion electron volts, or less than 1% of the energy in the head-on case. Figure 3.13 illustrates the basic differences between fixed target and collider operations.

A Brief History of CERN

One doesn't need to be much of a history buff to know that the early 1940s was a horrible time for Europe. The jackbooted shadow of storm troopers darkened most of the continent until 1944, when the largest amphibian invasion in history led to the desperate hedgerow fighting of Normandy. An endless series of trains rattled eastward toward camps, carrying their grim and tragic cargo and searing the camps' names into our collective psyche. Wave after wave of British bombers rained down fire on Hamburg, creating such a maelstrom that the air itself seemed to burn, reducing the city to ashes and killing 43,000 souls overnight. Fighters above Britain banked and rolled violently in a dance of death. The Red Army reduced Berlin to rubble. Europe was convulsed in a war so terrible that the destruction and suffering has not been matched before or since.

This consuming conflict ended in Europe in May of 1945, after which the victorious Allies conducted the Nuremburg trials, accusing, convicting, and condemning to death the Axis leaders for crimes against humanity—crimes so heinous that they needed a new name. The continent had been torn asunder.

Thus it remains astonishing to me that just a mere four years later, in 1949, Nobel Prize-winning physicist Louis de Broglie could propose a new pan-European physics laboratory in which all of the major European powers (victors and defeated alike) would participate. Just three years after that, eleven European governments agreed to create a provisional Conseil Européen pour la Recherche Nucléaire, or CERN. With the ratification by the member nations of the treaty setting up the organization on September 29, 1954, the provisional status disappeared and the European Organization for Nuclear Research, as it is known in English, came into being, although the CERN acronym was retained. The signatories to the treaty were Belgium, Denmark, France, the Federal Republic of Germany, Greece, Italy, the Netherlands, Norway, Sweden, Switzerland, the United Kingdom, and Yugoslavia, one more than the original eleven. Less than a decade after the bombs stopped falling, at least the scientific European community was healing. Perhaps even most surprising was the fact that the voting public in the signatory nations went along.

In 1952, Geneva, Switzerland, was chosen as the site for the new physics laboratory and, in 1957, CERN's first particle accelerator came online, a 600 million electron volt synchrocyclotron. This initial accelerator operated with energy about ten thousand times lower than today's LHC.

An accelerator of that energy was noteworthy more for its mere existence, initiating as it did CERN's decades-long presence (so far) as a juggernaut on the world's stage of particle accelerator laboratories. However, in 1959 CERN catapulted itself to the top of the particle physics world, with the turn-on of the

28 billion electron volt Proton Synchrotron (PS), for a short time the world's highest energy accelerator and still operating today as part of the LHC chain of accelerators.

While CERN's early accelerators and experiments were of the fixed target variety, in 1965 the world's first proton-proton collider was proposed. Commissioned in 1971, the Intersecting Storage Rings (ISR) Collider had an available collision energy of 62 billion electron volts. The ISR didn't have detectors of the type found at the LHC, so it was much more of an accelerator triumph than a physics one. For instance, it held the world record for the brightness of the beams at the collision points until 2004, when America's Fermilab Tevatron finally surpassed it—fully twenty years after the ISR's shutdown in 1984.

The year 1971 included a proposal for a new accelerator at CERN, the 4.2 mile (7 kilometer) circumference Super Proton Synchrotron (SPS). The SPS began operation in 1976. Its design energy was 300 billion (3×10^{11}) electron volts, although CERN's clever accelerator scientists were eventually able to run it at 500 billion electron volts of energy, or 7% the energy of today's LHC. The SPS project came in ahead of schedule and under budget and still plays an important role in the LHC complex.

While the SPS now operates as one of the accelerators in the LHC chain, it briefly was reconfigured to run in a bold, new way. The SPS was turned into a colliding ring. This was not all that innovative, after all the ISR had already done that. No, the innovation was that these colliding beams would consist of protons (nothing new there) and *antimatter* protons! While America's decommissioned Fermilab Tevatron holds the record as the world's best proton/antiproton collider, it wasn't the first, nor is it at all clear that the Fermilab accelerator would have turned on as cleanly as it did without the Super Proton-Antiproton Synchrotron ($Sp\bar{p}S$) leading the way.

The next noteworthy date in our whirlwind trip through CERN's history is 1981, when the 17 mile (27 kilometer) circumference Large Electron Positron (LEP) was approved. It was eventually designed to accelerate electrons and antimatter electrons (positrons) to an energy that was precisely selected to produce enormous numbers of Z particles. (You may recall from chapter 2 that the Z particle is a carrier of the weak force.)

In 1989, the LEP accelerator turned on and performed simply brilliantly. The measurements performed by the four competing experiments on the properties of Z particles may never be surpassed. My personal favorite LEP measurement is the one that is generally interpreted as proving that there are three—and only three—particle generations. (Remember that a generation is one of the carbon copies of subatomic particles, of which the up and down quark and the electron and electron neutrino are the first.) Technically, the measurement showed that there were only three light neutrinos, which leaves open the possibility of additional generations as long as they have heavy neutrinos. Either way, this measurement is an important clue to the particle puzzle. I just wish I could figure out what it is telling me. I say that a lot, but the universe is trying to tell us something profound. When someone figures it out, I'm going to say, "Well duh! I wish I had thought of that!"

A little side note illustrates the caliber of the effort that went into the accelerator's construction and operation. In order to understand their data, the CERN scientists needed to set the accelerator to a precise collision energy. After a while, during which time they commissioned the accelerator and experiments, they saw a peculiar thing. The energy of the beam varied over the course of a day. There appeared to be two cycles per day, with the time between cycles being about 12.5 hours and the time of day during which the maximum and minimum deviation in energy occurred about an hour later each day. Very peculiar. Lots of equipment were checked, heads were scratched and . . . um . . . colorful . . . language was uttered. (And, in the polyglot environment of CERN, this can be an entertaining thing to experience. Some languages really do have a flair for invective.) Eventually an innovative thinker had the long-sought "Aha!" moment. It turned out that the mysterious variation was caused by the effect of the lunar tides on the Earth. The force of the moon's gravity causes the surface of the Earth to flex by about a foot (30 centimeters). This had the effect of changing the radius of the 17 mile (27 kilometer) circumference LEP accelerator by about one twentieth of an inch (one millimeter). This works out to be a change of 0.00001%, and it was noticeable. Wow.

The LEP experiments made hundreds of precise measurements and searched for new phenomena, including looking for the top quark before losing out to Fermilab in 1995. Before the LEP accelerator was decommissioned in November of 2000, it was run at even higher energies, eventually more than doubling the energy of its original design. The idea was that if you're going to turn off an accelerator soon, you have little to lose if you damage the equipment.

In its last few months, the LEP accelerator delivered data that seemed to indicate that perhaps its experiments had discovered the Higgs boson. While these data are no longer believed to have supported that idea, at least LEP went down swinging. Its four experiments' results still play a valuable role in our current understanding of the Higgs boson.

The LEP accelerator was decommissioned in 2000 for a reason very important to readers of this book. This reason is because the LHC now inhabits the LEP tunnel. Out with the old and in with the new as they say, and a new era has begun.

Long before the LEP accelerator was decommissioned, the CERN governing council decided that the LHC in the LEP tunnel was the future of European particle physics. In December of 1991, this fateful decision signaled the eventual death knell of the LEP accelerator. The construction approval in 1994 sealed its fate. With the writing on the wall being rather apparent, the four LEP experiments put the remaining time to excellent use.

The period of time between 2000 and 2005 was marked by feverish activity. The LEP accelerator and all four experiments needed to be dismantled and removed. The LHC components needed to be assembled, with the first of the bending magnets installed in 2005. Final magnet production occurred in 2006, and the final bending magnet was lowered into place on April 26, 2007.

With our quick journey through CERN's history from the laboratory's beginning to the LHC era now complete, we finally come to the point of this chapter: a discussion of the LHC accelerator complex.

Nuts and Bolts

CERN is a fairly small site on the Franco-Swiss border. Most of CERN's accelerators are contained within its perimeter. However, this is not true for the LHC itself. The LHC breaks free of the site and swoops in a large circle for miles (kilometers) through the French and Swiss countryside. Well, to say it passes through the countryside is misleading. It actually passes under the countryside, on average about 300 feet (a hundred meters) underground. The actual depth varies from 150–450 feet (50 to 150 meters), depending on the location's proximity to the foothills of the Jura Mountains. Thus the LHC accelerator is invisible to the surface dwellers. People living above the ring pass their lives in blissful ignorance of the frantic dance of protons circling under their feet. Figure 3.14 shows a bird's-eye view of the area surrounding CERN, while figure 3.15 shows more clearly the subterranean nature of the LHC ring.

The LHC complex consists of five distinct accelerators and some equipment preceding the first "real" accelerator (figure 3.16). Prior to actual acceleration, one must obtain protons. One does that by taking ordinary hydrogen, which consists of a proton and electron, and stripping the electrons off. This is done by means of the Duoplasmatron source, which seems to have stolen its name from 1930s pulp science fiction. The Duoplasmatron source provides protons with an energy of a hundred thousand electron volts (1×10^5 eV). This meager energy works out to be about 1.5% the speed of light, but it is still about 2,700 miles (4,300 kilometers) per second.

The first "real" accelerator a proton encounters in its journey is Linac 2. (Linac 1 was CERN's first linear accelerator, or linac, and was decommissioned in 1992.) A linac is a straight-line accelerator, consisting of electric fields all pushing in the same direction. A proton enters the 256-feet-long (80 meter) linac

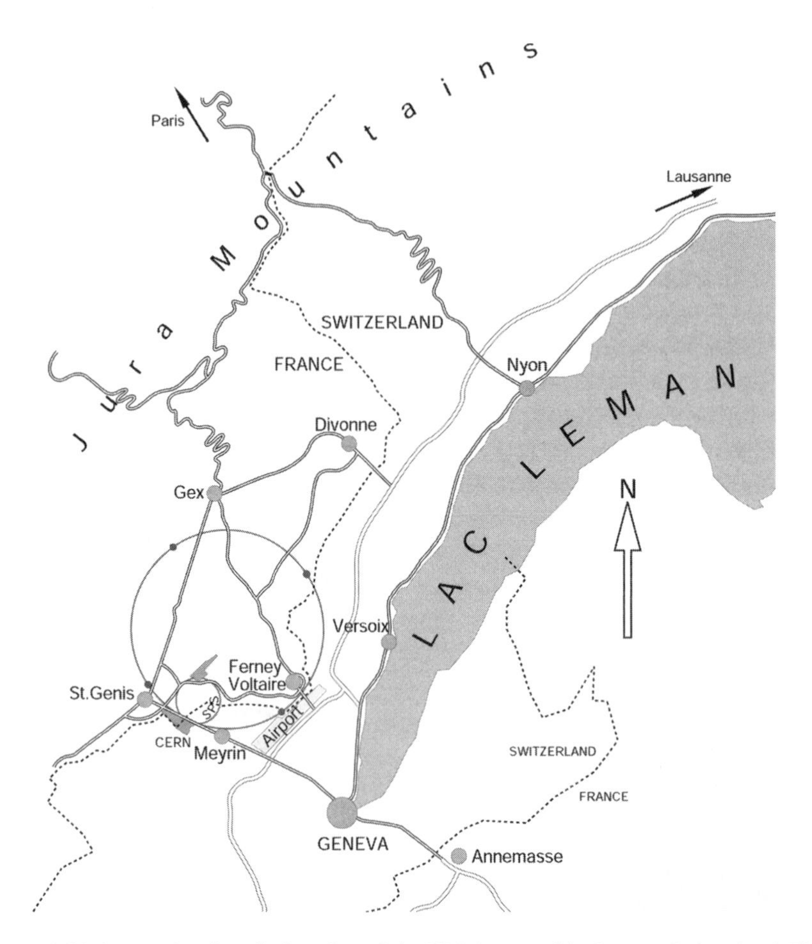

Figure 3.14. A map showing the location of the LHC, just outside Geneva, Switzerland. The 17-mile-long (27 kilometer) ring spans the Swiss and French border. The bodies of water near the collider are Lake Leman and the Rhone River (not shown). The outer ring shows the LHC and the smaller ring between St. Genis and Ferney Voltaire shows the location of an earlier accelerator, the Super Proton Synchrotron, or SPS. Figure courtesy of CERN.

with essentially no energy and leaves it with an energy of 50 million electron volts (5×10^7 eV). Fifty million of anything sounds like a lot, but it is far less than the LHC's ultimate energy of 7 trillion electron volts (7×10^{12} eV). In fact, a proton leaving the linac has about a millionth of the proton's final energy. Even so, the velocity of a proton leaving the linac is traveling 31% of the speed of light.

As the proton leaves the linac, it is guided into the first circular accelerator, the Proton Synchrotron Booster (PSB), a ring with a circumference of about

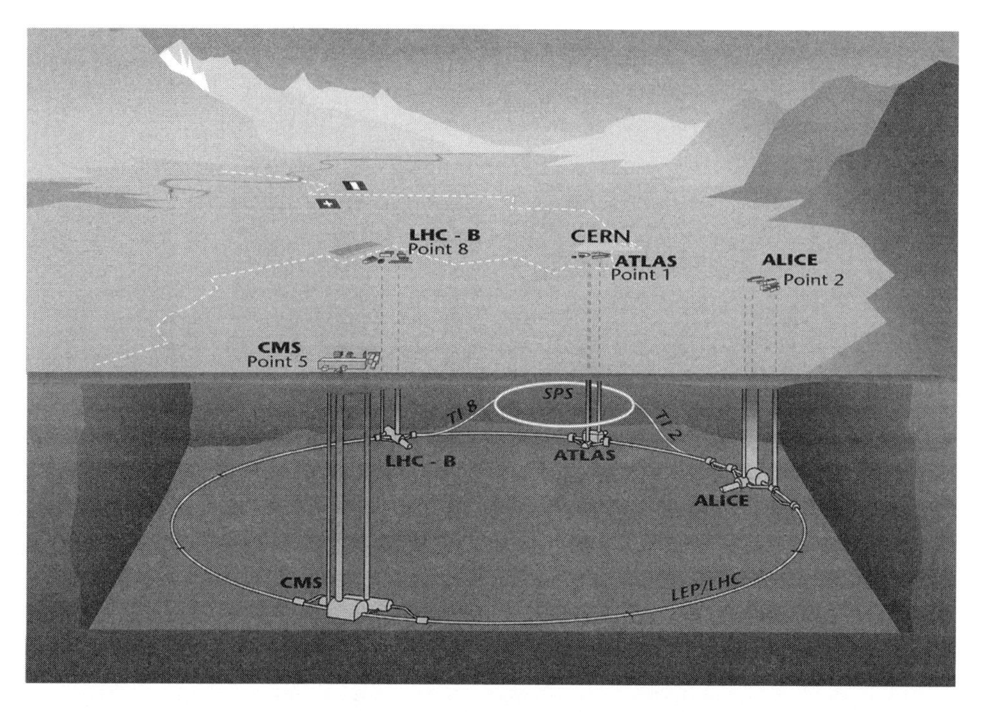

Figure 3.15. This drawing depicts the LHC accelerator complex, including the relative locations of the various experiments. This figure has very different vertical and horizontal scales, as the experiments are about 300 feet (100 meters) below the Earth's surface, and the ring is 17 miles (27 kilometers) around. The top part of the image shows what a person could see of the LHC complex aboveground and the lower part shows the elements that are deep underground. LEP refers to the old accelerator, Large Electron Positron, which used the space before the LHC was built. Points refer to eight spots along the ring at which collisions can occur. Figure courtesy of CERN.

500 feet (157 meters). The PSB increases the proton's energy from the linac's 50 million electron volts to 1.4 billion electron volts (1.4×10^9 eV). The proton's velocity leaving the PSB is 91.6% the speed of light.

The next accelerator is the Proton Synchrotron (PS). While the PS was once the world's highest energy accelerator, it is now but a way station en route to the LHC. The PS is a ring with a 2,010 feet (630 meter) circumference, and it raises the energy of the proton to 25 billion electron volts (2.5×10^{10} eV). The proton velocity leaving the PS is 99.93% the speed of light.

The SPS is the penultimate accelerator. It was once the site of the discovery of the W and Z particles (discussed in the first chapter). This fourth accelerator has a circumference of 20,160 feet (6,300 meters) and raises the energy of the

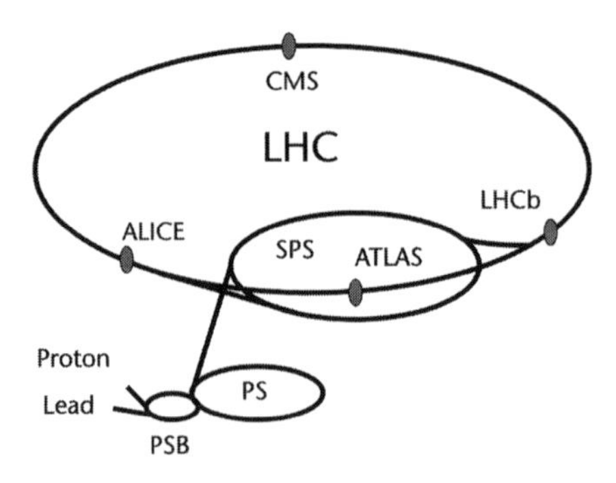

Figure 3.16. The Large Hadron Collider accelerator complex actually consists of five distinct accelerators (Linac, PSB, PS, SPS, and LHC), and this figure gives a sense of the relative accelerators' dimensions. The gray-colored ovals discussed show the location of the big LHC experiments, which are described in detail along with other components of the LHC in chapter 4.

proton to 450 billion electron volts (4.5×10^{11} eV). These protons, now traveling at the astonishing speed of 99.9998% the speed of light, are now ready for injection into the LHC.

The fifth and final accelerator is the LHC itself. The LHC accepts the protons from the SPS and is designed to increase their energy to the full 7 trillion electron volts (7×10^{12} eV). This is 100% of the design energy and brings the protons' velocity to 99.9999991% the speed of light. This is so fast that a person sitting in one of CERN's laboratories will see a photon only moving 8.5 feet (2.7 meters) per second faster than the LHC's protons. You'll note that the velocity in this final accelerator is not much faster than it was in the PS, although the energy is about 300 times greater. This is a consequence of Einstein's theory of special relativity, which describes how things move at high velocities, but it still seems odd to many readers.

In all of the circular accelerators except for one, the beams orbit either in a clockwise or counterclockwise manner. However, the LHC consists of two beams, one orbiting in each direction. In figure 3.16, we see that there are two different points at which protons can be transferred from the SPS to the LHC, enabling a clockwise-only accelerator to load the LHC with protons in both directions. This is just one of many clever engineering features embodied in this intricate equipment. The transfer lines were built in various locations in the area once claimed by the former Soviet Union. Shipping the transfer lines from their final assembly site in Novosibirsk to CERN was an epic journey.

All five accelerators have electric fields formed by the radio frequency (RF) cavities discussed earlier in the chapter. Further, all of the accelerators except for

Linac 2 have bending magnets. Rather than discussing details about all of the accelerators, let's focus on the LHC.

Describing the LHC really involves a long list of numbers. Every counter-rotating proton beam is accelerated by eight RF cavities. Each RF cavity adds two million electron volts of energy, thus giving each beam an additional 16 million electron volts of energy on each orbit. So, in the absence of other considerations, the proton beam could be fully accelerated in 400,000 orbits. Given that the beam makes a little over ten thousand orbits per second, the acceleration phase could take as few as 40 seconds.

The actual acceleration time is much longer, more like 20 minutes. The acceleration period is much longer in part because the electric current in the various magnets must be increased to strengthen the magnetic fields as the beam energy is increased. In order to keep the beams inside the beam pipe—which is a tube a couple of inches (a few centimeters) wide and 17 miles (27 kilometers) long—the magnetic fields must be constantly adjusted to tame the beams' increased energy.

The bending magnets are capable of running with a magnetic field of 8.3 Tesla, or about 170 thousand times more powerful than the Earth's magnetic field. The Earth's magnetic field is what makes a Boy Scout's compass such a useful tool (or Girl Scout's, I suppose, but I was never one of those). This incredible magnetic field is achieved by running 11,800 amperes of current through the wire of the bending magnets. The power distribution panel for my house is rated for a maximum of 100 amperes, so each bending magnet will use the current necessary to power over 120 ordinary houses. The entire string of 1,232 bending magnets consumes the electric current that could supply the needs of a small city, consisting of 150,000 houses. When the entire LHC complex is considered, the total power consumption is about 120 million watts of electrical power at peak demand.

With so much electrical energy stored in the magnets, it is natural to wonder how much that works out to be. The total energy stored is an astounding 11 billion joules. To give some perspective, that is about the amount of energy stored in two and a half tons of TNT (although in the case of the LHC it is spread out over 17 miles). This is equivalent to a 400-ton 747 jet airplane hitting the ground at about twice the speed of sound.

Each bending magnet is about 50 feet (15 meters) long and weighs about 35 tons. The wires that form the electromagnets are made of a superconducting niobium-titanium alloy and resemble the kind of wire that powers household appliances. Cut the power cable to a table lamp, and you'll see that each cable is made of smaller strands, twisted like a rope. In the case of the LHC's superconducting wire, it is made of strands, which are in turn made of filaments. The length of strands that make up all the LHC's bending magnets would go around the Earth's equator about seven times. The length of the filament involved would stretch five times from the Earth to the sun and back, with enough left over for a few trips back and forth to the moon.

Getting these magnets to the proper temperature requires an extraordinary amount of coolant. A total of 10,800 tons of liquid nitrogen is needed, followed by 120 tons of liquid helium. The entire process takes about 6 weeks, during which the 37,000 tons of magnets are cooled from room temperature to their desired temperature of –456°F (–271°C).

While it takes a little more than 4 minutes to transfer enough protons from the SPS to the LHC and about 20 minutes to accelerate them to their maximum energy, the beams are then left to collide in the detectors for about 10 hours. In order for the beam to last so long, the pipes through which the beams circulate must be under a superb vacuum. If a vacuum were not in the pipe, the beams would interact with the molecules of air in the pipe and immediately disappear. The vacuum in the beam pipe is 10 trillion times rarer than ordinary air and the pressure is ten times lower than that on the surface of the moon (10^{-13} atmospheres for the technical crowd). This is among the best vacuums achieved on Earth. Perhaps even more impressive is the total volume that needs to be a vacuum. This volume is about 220,000 cubic feet (6,500 cubic meters), or about equivalent to pumping all of the air out of one of Europe's many majestic cathedrals.

The last topic to cover is the structure of the LHC beam itself. As mentioned before in our discussion of surfers, the beam is composed of bunches of protons, with each bunch separated by no less than 25 feet (7.8 meters). You may recall that this is the distance between adjacent waves in the accelerating electric field. Of course, it is possible for adjacent bunches to be farther apart than 25 feet. If not every wave is filled with protons, adjacent bunches could be 50, 75, 100, and other increments of 25 feet apart, depending on how many "waves" are skipped. While the LHC design calls for 25 feet separation, the 2010 to 2012 running used a separation of 50 feet. During the next running period, we expect the accelerator to perform as designed, but this isn't something we'll know until the machine is turned on.

So let's look at an individual bunch of particles. One bunch in the LHC includes about 100 billion (10^{11}) protons. The actual shape of each bunch has a passing resemblance to a stick of uncooked spaghetti, although the length is about a foot (30 centimeters) long and the width is a little less than a thousandth of an inch (0.03 millimeters). At full design, there will be 2,808 bunches of protons orbiting in each direction and aimed and focused to collide at four points around the LHC's perimeter. So, except for the fact that the actual beam width is about a hundred times smaller than a piece of spaghetti, you can get a pretty good visualization of the LHC's beam as about 3,000 pieces of uncooked spaghetti, each separated by 25 feet (7.8 meters), orbiting at 99.999999% the speed of light, in an orbit that is 17 miles (27 kilometers) around. If you do the math, you find that about 3,800 lengths of 25 feet are needed fill up the whole circumference of the orbit. So if there are 2,808 bunches, the entire accelerator is not filled. In fact, you have a concentrated group of bunches, each separated by 25 feet (7.8 meters), followed by a longish gap with nothing in it. This longish gap has many uses, although "longish" is a relative term and is in the ballpark of a millionth of a second. This exceedingly brief time during which there are no protons in the detectors is used for the detectors to recover and reset themselves.

The energy stored in these beams is enormous, although only about 3% that stored in the magnets. However, this beam must pass through the center of the various detectors in the center of the experiments spaced around the ring. The equipment in the experiments that is near the beam is extremely delicate and consequently extreme care is taken to be able to nearly instantaneously dump the entire beam into a large absorber if there is the slightest indication that the accelerator operators are losing control of the beam. The beam carries so much energy that, were it not controlled carefully, it could easily destroy the heart of the particle detectors. To give you an idea of how much energy we're talking about, it's about 350 million joules. That's as much energy as a 400-ton commuter train traveling at 100 miles per hour or enough to melt a half a ton of copper. To protect the equipment, this energy is diverted to a stack of graphite absorbers in under a thousandth of a second. That's like absorbing the energy of a relatively large military conventional (i.e., non-nuclear) air-dropped bomb.

These two counter-rotating beams will each circle the LHC accelerator ring a little over ten thousand times per second. When the LHC is operating at full potential, something like 800 million collisions per second will occur in each of the detectors, although most collisions will occur through well-known and therefore not very exciting physical processes. These beams will collide for 10 or maybe as much as 20 hours, during which time they will travel about 10 billion miles (16 billion kilometers). That's about like traveling to Neptune and back, all the while circulating in a pipe a few inches (several centimeters) wide. No matter how you look at it, the LHC is an extraordinary technological marvel.

Smashing Lead

Before we close this chapter, we need to discuss one additional thing. We've focused predominantly on the case where the LHC is colliding beams of opposing protons. But the LHC is not a one-trick pony. The LHC is also designed to be able to accelerate heavy ions, which are atomic nuclei stripped of all their electrons.

While the LHC can accelerate many different heavy ions, its design is optimized to accelerate lead nuclei. Lead consists of eighty-two protons and 126 neutrons. The process whereby lead is accelerated is similar to the proton case, so we'll only discuss the main differences.

A pure sample of lead is heated to about a 1000°F (540 C). An electric current is passed through the lead to knock some of the electrons off the nuclei. Each lead atom contains eighty-two electrons, and the current can typically knock about thirty electrons off but usually not many more. The lead nuclei are accelerated through a different linac, called Linac 3, and the lead beam is passed through a thin carbon target, which knocks off another twenty electrons or so. In order to accumulate enough lead to make enough interesting collisions, the beam is guided into an accelerator called the Low Energy Ion Ring (LEIR).

When enough lead has been stored in the LEIR, the lead is then transferred to the PS accelerator, which accelerates the lead and passes it through another target, which knocks off the remaining thirty or so electrons. The lead nuclei, now stripped of all their electrons, are passed through the SPS into the LHC. In the LHC, the lead nuclei are accelerated to 2.8 trillion electron volts per nucleon. Recall that when the LHC is accelerating protons, each proton carries 7 trillion electron volts. It first seems like the LHC lead beams are lower in energy than the proton beams. However, recall that each lead nucleus contains a total of 208 protons and neutrons. This means that the total beam energy per lead nuclei is about 575 trillion electron volts, resulting in the most violent, large-volume collisions ever recorded. The LHC is designed to run in the mode in which heavy ions are accelerated about a month each year.

LHC Meets Literature

If you've ever read the book *Angels and Demons*, you'll recall that it talks of a plot to take antimatter from CERN to blow up the Vatican. Much of the description of CERN is false. There are no hypersonic planes and no wind tunnels. Ivy covered walls are in short supply, although Frisbee matches on the lawn are a frequent sight.

One plot device in the novel is the retina scanner, which is a security mechanism used to ensure that only two people can enter a particular laboratory. The reality is similar. At CERN, if you wish to enter the tunnel housing the LHC or any of the caverns in which the various experiments are located, one of the procedures involves going inside a compartment the size of a telephone booth (for those of you who actually remember telephone booths) and putting your eye up to a camera. Your iris (not retina) is photographed (sorry, no dramatic light crossing your eye) and compared with information stored on your key card and a database of people who are authorized to enter the restricted area. If you are, the inner gate opens. If not, no dice. There are also scales to ensure that only one person is inside the booth at a time, as well as cameras, keys, and other safety features. The fundamental idea is to ensure that only authorized people can enter and that the safety personnel know at any time exactly who is inside the secured area.

Future

The LHC is by no means a static facility. The accelerator was supposed to run at 14 trillion electron volts and with a luminosity (i.e., beam brightness) of 10^{34} in the obscure units of $cm^{-2}\ s^{-1}$. Let's just call that nominal brightness to be 100%. We can then talk about the brightness relative to that.

For reasons that we will look at in far more detail in chapter 5, during 2010 the LHC ran at half the design energy (7 trillion electron volts) and a maximum brightness of 2%. In 2011, the energy was unchanged, but the beam brightness rose steadily from about 1% to 40% of design brightness. In 2012, the beam energy was raised to 8 trillion electron volts (still not much above 50% of the design), while the beam brightness was about 60 to 70% of what was originally planned.

During its next operational phase, the LHC is expected to run at near 13 trillion electron volts and at about 100% the expected beam brightness. But then things start to go up from there. After running through 2017, the LHC will undergo a second long shutdown (the first being from 2013 to late 2014/early 2015). When the machine resumes operation, the plan is to be at full energy and 200% of the design brightness. In 2022 or so, another shutdown will occur, and the LHC will be upgraded to something like 500% of nominal beam brightness. While you should not take any of these numbers or dates as written in stone, they illustrate that the performance of an accelerator is a moving target, changing as the experience and cleverness of the engineers and operators allow.

I talk of the beam brightness, but that is not the entire story. The amount of data we gather is a mix of both how bright the beams are and how long the beams are colliding. If someone gave you a hundred dollars per second for one second, you'd get a hundred dollars. If someone gave you a dollar a second, that is only 1% of the rate. Yet if they gave you one dollar a second for 200 seconds, you'd have twice as much money as you'd have gotten with the high-rate, short duration generosity.

The amount of recorded data is expected to grow year after year. Again, the unit of delivered beam is particularly obscure, specifically 1/(area), where area has the ordinary meaning of length times width. In order to give a sense of scale, let's just call the amount of data recorded by either ATLAS or CMS in 2011 as a single unit of data, meaning we say the 2011 data set is 1. In 2010, the delivered data were about 0.01, while in 2012 the recorded data were at about 4, or four times the amount in 2011. After the first long shutdown, the LHC is expected to run until 10 or 20 units of data are recorded. After the second shutdown, the data will be recorded at a rate of 15 or 20 per year and will add up to maybe 50 or so units. After the third shutdown, the LHC is expected to run until perhaps 500 units of data are recorded. This running period will begin after about 2025 or so. In 2030 or so, it is likely that the LHC will have thoroughly explored what is available, and a future accelerator will be under way.

While the details of the more distant accelerator upgrades are necessarily evolving, one near-term upgrade is expected and that is the replacement of the linear accelerator part of the accelerator (Linac 2) with a new one (Linac 4). This upgraded accelerator will inject beam into the PSB at three times the energy of the existing facility. Linac 4 should be ready to accelerate protons when this book goes to press, and negative hydrogen ions (protons with two electrons) by late 2016.

By any measure, the LHC accelerator is a highly complex instrument. It is intended to concentrate an unprecedented amount of energy into incredibly tiny volumes. However, no matter how impressive a technical achievement the LHC accelerator is, if the collisions are not recorded, the whole exercise is pointless. In the next chapter, we will discuss how these particle collisions can be recorded.

4

INCREDIBLE DETECTORS

While the effort put into accelerating particles described in the previous chapter is necessary to understand the microworld, it is not sufficient. One must also record the collisions between protons by taking what amounts to fast and high-tech photographs. By recording and reviewing millions and indeed billions of these collisions, you can get a good idea of the kinds of things that can happen in collisions between protons, from the common to the rare. Finally, by understanding why the common things are common and the rare things are rare, you can learn a great deal about the behavior of matter and energy under extreme conditions and even about the birth of the universe itself.

So, just how does one record the collisions caused by a large particle accelerator such as the LHC? You need huge detection equipment, weighing thousands or tens of thousands of tons. In a collision likely to occur in the LHC, two particles enter the collision (the protons) and lots (say somewhere in the neighborhood of 10 to 500) of particles come out. The story of each collision is etched in the trajectories and the identities of the outgoing particles and can be displayed to scientists in an image called an *event display*. Because the actual collision occurs in such a mind-bogglingly short time, the collision itself is usually hidden from us. It's only by looking at the debris of the collision that we can understand the important stuff. Understanding particle collisions is essentially a study in forensics.

You can understand the mind-set of an experimental physicist if you pretend to be a bomb-squad investigator. Bomb investigators do not generally understand the details of the explosion by being close by when it occurs . . . at least not if they want a second assignment. No, bomb investigators understand the explosion by studying its effects on its surroundings. By studying scorch marks, total damage, amount of debris, and how deeply the shrapnel penetrates well-understood materials, the expert can get a good idea as to what happened. Chemical analysis adds to the story.

Table 4.1. A comparison of the various major detectors at the LHC

	ATLAS	CMS	ALICE	LHCb
Weight (tons)	7,800	14,000	11,000	4,700
Height (ft)	70	48	51	32
Length (ft)	147	77	83	64
Price (million $)	460	460	125	63
Magnet strength (relative to Earth)	40,000	80,000	10,000	40,000
Energy (TeV)	14	14	1,150	14
Brightness (relative to ALICE)	10,000,000	10,000,000	1	100,000
# collisions/second	800,000,000	800,000,000	8,000	40,000,000

Note: TeV = one trillion electron volts

Similarly, particle physicists study their collisions by surrounding the collision point with a detector of well-known and carefully selected composition. By seeing how the particles leaving the collision interact with the detector, their energy, trajectory, and point of origin can be inferred. The right detector will reveal at least some of the particles' identities. This information can be brought together like a jigsaw puzzle, with each piece of information neatly interlocking and revealing the true picture of the initial collision.

In this chapter, we look at the simple building blocks and technical considerations involved in the design of any modern detector. After that, we examine details about each of the major detectors arrayed around the LHC. We will concentrate on the ATLAS and CMS detectors, with fewer details given for the ALICE and LHCb detectors (see table 4.1 for basic information about these four detectors). The TOTEM and LHCf detectors are extensions of ATLAS and CMS and will be mentioned only in passing.

Before we discuss technology and techniques, we must spend a moment talking about the kinds of particles we need to be able to detect. There are literally hundreds of kinds; however we don't need to know about all of them to understand the most important points. The handful of particles we need to know about are these: the electron, the photon, the muon, the neutrino, and a class of particles called *hadrons*. Electrons and photons are relatively familiar, appearing as they do in the world of human experience: electrons in electricity and photons in light. Electrons and photons do not penetrate deeply within a detector.

Muons and neutrinos are less familiar but were introduced in chapter 1. Muons are basically heavy electrons, although they interact very little with the detector and usually pass through it, leaving behind only a small fraction of their energy. Neutrinos have no electric charge and nearly no mass and experience only the weak force; they pass through a detector without interacting at all. In essence, they are not seen in a detector, and their presence is known only by their absence.

Hadrons are a class of particles that contain quarks within them. The protons and neutrons are the most familiar hadrons, although they are relatively rare in the debris of particle collisions. Pions, short for pi meson—a particle consisting of a quark and an antiquark—are the most common type of hadrons in a particle collision. They can be treated in many ways as if they were light protons, having only about 15% of their mass. The manner in which hadrons interact with matter is midway between those mentioned for electrons and those for muons. Hadrons penetrate more deeply than electrons and photons but not nearly as deeply as muons and neutrinos. The different ways in which these particles interact with matter play an important role in revealing their identity.

Identifying the point of origin of a particle is often quite important. This is because sometimes rare particles are made that live for a long time . . . several trillionths of a second. This seems very short, but highly energetic particles with this lifetime live long enough to travel as much as an inch or so (2.5 centimeters) before decaying (although most particles don't make it that far). Since particles that live this long frequently occur in rare physical processes, we'd like to pinpoint when these kinds of particles are made. Typically, one identifies such events when the trajectory of particles in the event is reconstructed and it becomes apparent that not all of them originated from the same point that the protons collided. When we project particles back to their point of common origin, we call this a *vertex*. Vertices that differ from the collision point are of interest to researchers. Figure 4.1 shows what the signature of a long-lived particle might look like.

Identifying Particles

While there are many clever techniques for discovering the identity of particles, we only need to know a few. The topics we will discuss are *magnetic bending*, *ionization*, *showering*, and the rather ominous-sounding duo, *transition radiation* and *Cerenkov radiation*. We'll introduce each of these ideas in turn.

Magnetic Bending

The first of the techniques, magnetic bending, is one we've encountered already. In the previous chapter, we discussed large circular accelerators, which you might recall consisted of a short acceleration region and a vast array of magnets whose sole purpose is to guide the protons in a circular path back for another acceleration phase. The crucial point here is that charged particles move in a circular path when they are being influenced by a magnetic field.

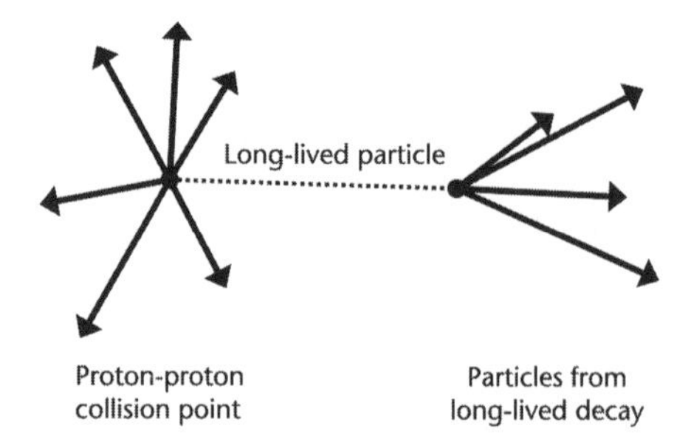

Figure 4.1. When two particles collide, you can identify the collision point by studying the trajectory of particles made in the interaction, which take the graphic form of event displays (for examples, see figures 4.26 and 5.7). These trajectories originate in a common point. However, occasionally some postcollision particles will appear to have a different origin. These seemingly peculiar particles actually are the signature of the decay of a long-lived particle created in the original collision. Because this particle was long-lived, it traveled an appreciable distance and subsequently decayed, resulting in a displaced origin for its decay products.

This fact can be exploited to help identify and measure the charged particles coming out of a collision. A circle is a simple geometric shape. The only thing that distinguishes different circles is their size. So, in order to be a useful technique for measuring particles, we have to be able to relate a circle's size (that is radius or circumference) to an important particle property. This turns out to be possible, and the important property is the particle's *momentum*. In our ordinary experience, momentum is related to the velocity of the object; the higher the velocity, the higher the momentum.

At the high energies involved in modern particle physics collisions, the correspondence between velocity and momentum doesn't hold, but it's still a valuable mental picture. However, at these high energies, momentum is more like energy. Because the term is more familiar, I will apologize to my physicist colleagues and use the word "energy" here.

Because the size of the circular path followed by a charged particle is related to the particle's energy, by measuring the size of the circle you have simultaneously determined the energy carried by the particle . . . the bigger the circle, the bigger the energy (figure 4.2).

Here is another interesting feature about the relationship between charge and the circular path: Particles with opposite electric charge (for example an electron and an antimatter electron) curve in opposite directions. If a positive particle moves counterclockwise, a negative particle will move clockwise.

Figure 4.3 shows an example of a relatively simple particle collision. In this example, five particles exit the collision. The particles have different energies

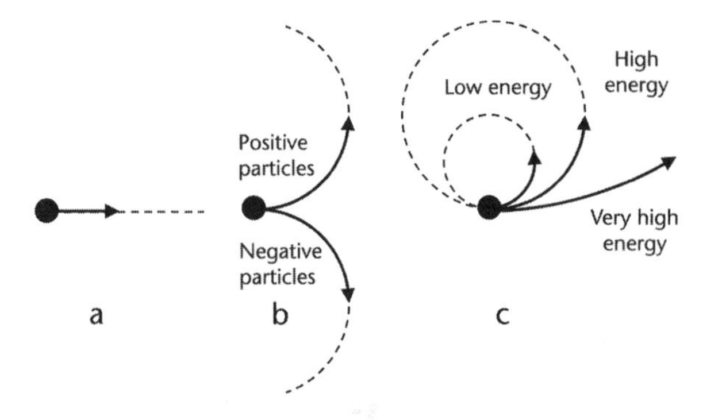

Figure 4.2. Magnetic fields alter the trajectory of particles, with the manner in which it alters the trajectory dependent on the type of the particle crossing the field, for example, when a magnetic field does not deflect a neutral particle (*a*), when a magnetic field deflects particles with electric charge, with particles of opposite charge being deflected in opposite directions (*b*), and when particles with low energy travel the circumference of a circle with small radius and higher energy particles follow circles of larger radii (*c*).

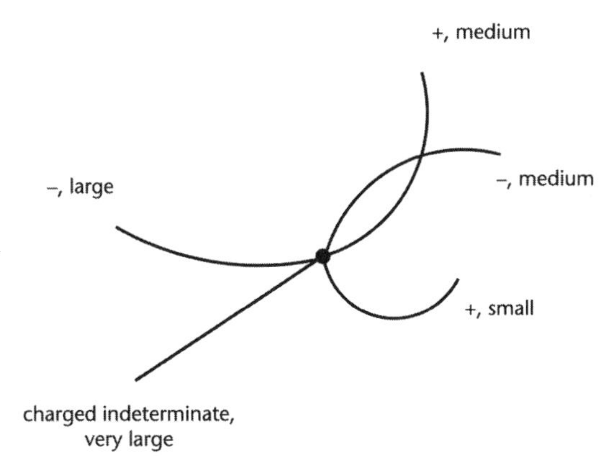

Figure 4.3. A magnetic field will alter the trajectory of charged particles passing through it. In the figure, the (+/–) signs indicate the type of electric charge and the words denote the amount of energy the particles carry. Note both the deflection direction and the curvature of the various tracks in relationship to their charge and energy.

and electric charges. From the clockwise and counterclockwise sense of the particles' motion, we see that we have two particles with positive electrical charge and two negative. The fifth particle has such a large energy that it is difficult to say if it is curving clockwise or counterclockwise. So, because of this ignorance, we are unable to say whether this particle is positive or negative. Further, this particular particle underscores an important limitation of this technique. For instance, if the particle is moving with so much energy that you can't tell whether it's moving clockwise or counterclockwise, then that essentially means that it is moving in a nearly straight line. Moving in a straight line means that it is moving along the circumference of a huge circle. This means that it has a lot of energy. The problem is that once you get to huge, you can't tell if the circle is huge, super-huge, or super-duper-huge. And, if you can't accurately measure the size of the circle, you can't accurately determine the particle's energy.

From this, we see that the magnetic bending technique works best when the energy of the particle isn't too big. This leads us to wonder how we can accurately measure the energy of highly energetic particles. This question is even more pressing, given that we know that the LHC is the highest energy accelerator ever built. Luckily, the technique called showering actually works better as the energy of a particle increases. We will describe this technique presently, but first we must discuss the second technique in our list: ionization.

Ionization

In our discussion of magnetic bending, we saw how the path of a particle related to the particle's energy. We didn't clarify exactly how we see the particle. For this, we need to talk about ionization.

When a charged particle passes through a chunk of material, it bounces into the atoms in the material. Unlike a bowling alley, in which the ball must physically hit a pin to knock it over, the charged particle is surrounded by an electric field. This electrical field extends far beyond the size of the charged particle itself. This electrical field can reach out and jiggle the atoms of the material through which it is passing.

Since this is a bit tricky to see, let's think up some analogies. If you take a magnet and move it near some iron nails, sometimes the nails will be attracted to the magnet, even though the magnet doesn't actually touch them. Or one might think of a big truck moving down a street covered by newspapers and Styrofoam cups. The wind from the truck's passage will move the debris around, even though the truck never physically hits it. So too it is with the electric field surrounding a particle carrying electric charge. As this effectively large object (the charged particle with its extended electric field) plows through material, it bounces into the material's atoms. With each bounce, the charged particle slows down just a little, like a bowling ball rolling through a room filled with pins. Fundamentally, that's all ionization is: a charged particle moves through material, bouncing into the material's atoms and slowing down in the process.

The next thing you need to know about ionization is that the amount of energy a particle loses is proportional to the distance of matter through which the particle travels. Say the particle loses one unit of energy after traveling through an inch (2.5 centimeters) of matter. Then if it travels through 3 inches (7.5 centimeters), it loses three units of energy. Fifteen inches (38 centimeters) means fifteen units of energy loss and so on. Conversely, if you measure the distance through which the charged particle travels, you know its energy. In our example above, if a particle travels through 100 inches (254 centimeters) of material and then stops, you know that it had 100 units of energy when it started.

Let's step back and take a look at what ionization means. To all intents and purposes, it's the same as slamming on the brakes of your car. The loss of energy due to ionization is effectively similar to the loss of the car's energy due to the friction between the tires and the road. And, just like a long skid mark means the car was moving quickly when you hit the brakes (i.e., it had a large initial energy), a charged particle penetrating deeply into matter means its initial energy was large.

Just how deeply can a particle, slowing only by ionization, penetrate into matter? Well, obviously that depends on the energy of the particle and the material through which it travels. Taking a relatively low energy particle (10 GeV for the technical types) in solid iron, a particle can travel about 25 feet (8 meters). Now given that the energy involved in an LHC collision is 14,000 GeV, particles with such a low energy will be very common. Particles with ten times as much energy will be pretty common as well. So these higher (but relatively common) energy particles would require a chunk of iron about a football field deep to stop them.

Given that modern particle physics experiments can't be that big—imagine a sphere of iron, 600 feet (190 meters) in diameter around a collision point, costing about 14 full years of the annual spending on physical science research by the U.S. government at 2011's price levels—there must be another solution or different technique we can use. This budget-saving technique is called *showering*.

Showering

While everything we've said about ionization is true, for some particles, it's not the entire story. Some particles will undergo additional types of interactions. The particles in question are electrons, photons, and hadrons (i.e., quark-containing particles). When these particles pass close to an atom, in addition to slowing down through ionization, they can actually split into two or more particles. For instance, if an electron passes close enough to a nucleus, it can kick off a photon: one particle in (electron) and two particles out (electron and photon). Similarly, when a photon comes close to the nucleus of an atom, it can disappear and be replaced by an electron and a positron: again one particle in and two out.

Here's a nifty thing. When a particle splits into two particles, they each get

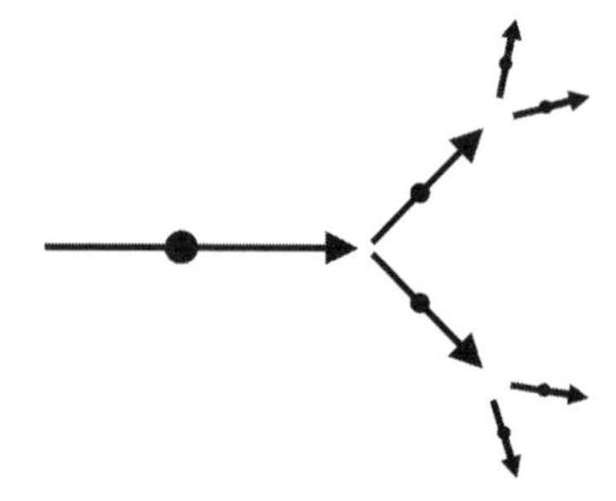

Figure 4.4. Showering is the simple process whereby one particle, shown as a large dot at the left of the figure, becomes two (or more). After the conversion, the "daughter" particles will share the energy of the "parent." Subsequent interactions further the process, turning a single, high-energy, particle into many, lower-energy ones.

half of the energy. So if the distance a particle can penetrate into matter is related to its energy, this splitting has converted one particle that can go a certain distance into two particles that can go half that distance.

To appreciate showering, you need to know something else. Once the one particle splits into two, well then, these "daughter" particles can also hit atoms and split. In this way, one particle can turn into two, then four, eight, sixteen, and so on. Indeed, it isn't at all unusual for one particle to "shower" into ten thousand (figure 4.4). With that increase in particle count comes a reduction in each particle's energy. Most important, showering vastly reduces the amount of material needed to fully absorb a particle and measure its energy.

The quark-carrying hadrons shower as well, although the details are a bit different. When all effects are taken into account, we find that hadron showers last longer and penetrate more deeply. Roughly speaking, the electromagnetic electron and photon will penetrate a few inches (10 or so centimeters) into a dense material like metal, while hadrons will penetrate a couple of feet (half a meter or a meter or so).

With the introduction of ionization and showering, we can get a first glimpse of how physicists start to distinguish between particle types. Let's take a simple two-component detector, with one section gaseous and in which ionization is measured and one section solid, in which showering occurs. (Don't worry about how the ionization is measured; we'll get to that later.) To show the essential points of how different particles are identified, let's consider five types: neutrinos, muons, photons, electrons, and hadrons. Neutrinos are electrically neutral and don't shower. Photons are electrically neutral and shower quickly. Electrons have electric charge and shower quickly. Hadrons can be neutral or

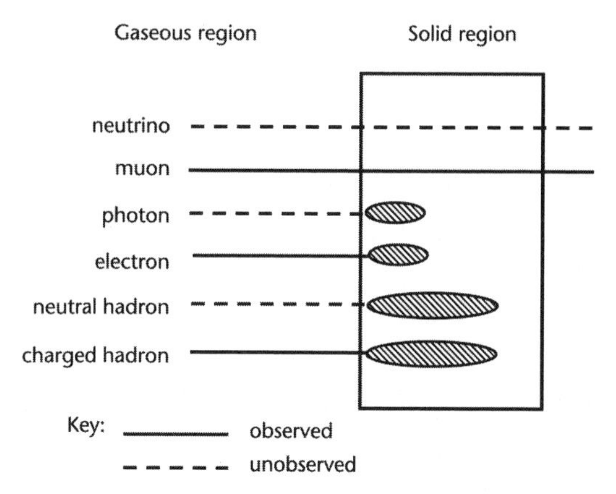

Figure 4.5. Various particles interact differently with matter. These differences allow scientists to identify the original particle. By observing whether the particle ionizes the gas region or showers in the solid one, we can fairly reliably identify the particles' nature. The ellipses indicate a shower.

electrically charged and shower slowly. Finally, muons have an electric charge but don't shower.

In figure 4.5, we see that electrically charged particles are observed in the gaseous form, but neutrally charged ones are not. In the solid state, particles shower as their nature dictates. By looking at the patterns in both detectors, the identity of the originating particles can be determined with considerable reliability.

That Pale Blue Glow

Before we get into the specifics of detectors, we need to introduce two additional useful effects: Cerenkov radiation and transition radiation. Most people with even the smallest science interest and training know that you can't go faster than the speed of light. (Although, judging from the crank letters and e-mails I constantly receive, this fact is not universally accepted.) Technically, the right thing to say is that you can't go faster than the speed of light in a vacuum. However, when light travels through a material, it moves more slowly. In fact, light travels through glass or Plexiglas about two-thirds the speed it has in a vacuum. This fact forms the basis for how lenses, prisms, and any number of optical phenomena work.

The fact that light travels more slowly when moving through glass leaves a loophole in that whole "faster than light" thing. This is because when light hits glass, it slows down immediately (and speeds back up when it leaves the glass).

However, the mechanisms that cause the slowing do not affect particles other than the photon. Thus we are left with the following situation: Suppose you had two high-energy particles, one electron and one photon, traveling alongside one another in a vacuum. No matter how high the energy carried by the electron, you'd see the photon pull ahead of the electron. If the electron was very high in energy, the photon might inch ahead, but ahead it would pull.

Now send these two particles into a slab of glass. The photon would immediately slow down to about two-thirds its initial speed, while the electron's speed would be essentially unchanged. Thus in glass, the electron can be faster than the photon! When this happens, an effect occurs that is similar to a sonic boom. A sonic boom occurs when an airplane moves through air faster than sound moves through the same air. Similarly, when an electrically charged particle travels through glass faster than light travels through the same glass, it gives off light. When this light is observed, you can be sure a highly energetic charged particle has passed through the glass. This light is called *Cerenkov radiation* or *Cerenkov light*.

We can merge the ideas of Cerenkov light and showering to create a powerful particle detection tool. We will come back to this idea later, but suppose you have a block of lead glass, which is used to make any high-end chandelier. Unlike ordinary glass, which is essentially sand that has been melted and cooled, lead glass consists substantially of lead. Recall that, when an electron comes close to an atom, it gives off a photon. Further recall that a photon, passing near an atom, will split into electrons and positrons. Finally, recall that these daughter particles can also pass near atoms, and the process will repeat itself. This is showering, as we learned earlier. However, in this case, the shower grows in the glass itself. This kind of detector doesn't consist of separate layers of high density metal that causes the shower and a separate detector medium. The metal is actually part of the glass. Since the electrons in the shower are very fast (and exceed the speed of light in glass), the electrons and positrons emit Cerenkov light, which can be collected and converted into electricity for further processing. So a chunk of lead glass and a high-tech electric eye can provide a way to measure the energy of electrons and photons.

Transition Radiation

The last technology we're going to describe is *transition radiation*. As its name suggests, this is radiation caused by a transition. Ah, but what transition? When a charged particle travels through a medium such as glass, it is surrounded by an electric field that is determined by its own electric charge and by that of the surrounding medium. Since the electric field depends in part on the medium

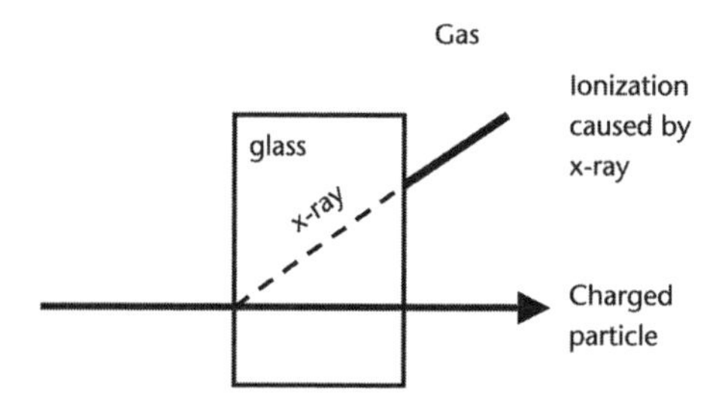

Figure 4.6. ***Transition radiation*** **occurs when a charged particle travels from one kind of material to another. When it crosses the boundary, a charged particle emits x-rays. These x-rays can ionize a gas and allow the x-rays to be detected.**

through which it travels, the electric field will change as the particle passes from one material to another (say glass to air or plastic to liquid). In the transition from one medium to another, an x-ray photon is emitted from the charged particle (figure 4.6). X-rays themselves are not seen, but they have enough energy to interact with the material and induce ionization.

If you carefully select the materials the particles pass through and their shapes, you can precisely locate where a charged particle has made the transition from one medium to another. Further, since transition radiation depends on a particle's velocity, this phenomenon can be used to distinguish fast particles (like the light electron) from slow particles (like the heavy, quark-carrying hadrons). The ATLAS and ALICE detectors at the LHC use this technology, and we'll describe how in more detail when we look at these detectors more fully later in the chapter.

Seen and Unseen

We now know several concepts important in particle physics detectors: magnetic bending, ionization, showering, Cerenkov radiation, and transition radiation. We also recall that we want to know as much about the particles coming out from a collision as possible, with special attention paid to their point of origin (usually the place where the collision occurred), their trajectory, electric charge, energy, and identity. It's time to bring these ideas together. Obviously there are as many different possible solutions to making the desired measurement with the available technologies as there are clever scientists and engineers. Accordingly, we will restrict the discussion to those choices made as part of the design of the various LHC-based detectors.

To understand the choices one might make, we first draw a simple diagram, with only a few kinds of particles. Like the earlier showering discussion, we will

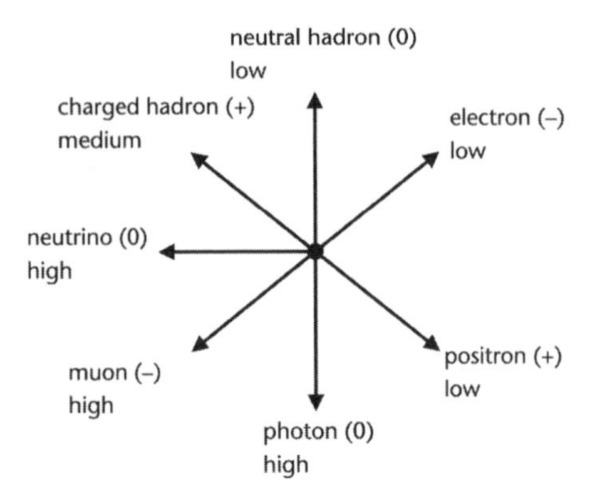

Figure 4.7. After a collision, particles exit the vertex, or point of common origin. Denoted here are the identity, electric charge, and energy of the various particles. The next few images show particles exiting the same collision but in each image a few more of the effects introduced in this chapter applied to it, culminating in what experimental physicists actually observe.

include an electron, a photon, a positron (an antimatter electron with the opposite electric charge as an electron), a muon, a neutrino, and electrically charged and neutral hadrons.

Figure 4.7 shows these particles, when you know everything about the particles you are studying. The identity and electric charge of each particle is given, as is how much energy each one carries. This is what physicists call the *truth level*. But, of course, our detector doesn't provide us with this perfect knowledge. In the following paragraphs, we are going to apply some of our techniques and get an idea of what physicists actually see when they look at particles passing through the detectors.

Recall that we want to know the electric charge and energy of the particles. One of our techniques we discussed was magnetic bending, which makes particles with electric charge move in a circular path. The size of the circular path is related to the energy the particle carries. Further, particles with positive electric charge curve in the opposite direction than negative particles.

So let's apply a magnetic field to the particles of figure 4.7 and see what effect it has (figure 4.8). The low-energy electron and positron are bent a lot. The positively charged hadron with medium energy is bent a middling amount, while the trajectory of the high energy muon is bent only minimally. The other particles, being electrically neutral, are not affected by the magnetic field.

While we've made our first steps toward capturing and understanding a particle scattering collision, there's just one obvious problem. We've not actually detected the passage of the particles. To do that, we need to dig into our bag of

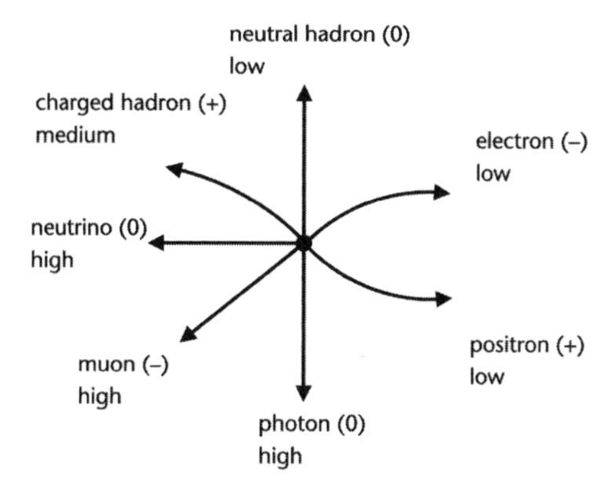

Figure 4.8. This is the same figure 4.7, but with the effect of a magnetic field imposed.

tricks. To view the particles' paths, we need to use ionization. Recall that ionization occurs when an electrically charged particle crosses through some type of material and interacts with the material's atoms. The effect of the particle's passage is then detected by various methods.

Typically in an ionization detector, you'd like to minimize the total amount of material through which the particles must pass. If you have too much, other effects we've not discussed (and won't) come into play, and things get complicated rather quickly. So to minimize what the particles must pass through, ionization detectors consist of many layers of material, separated by a low-density material, such as a gas or a vacuum.

In our simple example, we surround the collision with a series of concentric circles of material. Figure 4.9 shows what happens when the detectors are added. Electrically charged particles ionize the material and their passage is recorded, while the neutral particles slip through unscathed. In the figure, the passage of each charged particle through the ionization detector is recorded by a little dot.

So far, we've been able to see electrically charged particles but not the neutral ones. To see the neutral particles, we need to add the next trick: showering. Recall that showers provide a way for particles to dump all of their energy rather quickly in a dense material. Electrons, positrons, and photons have short showers, while hadronic particles have longer showers.

Shower detectors are generally comprised of thick slabs of dense material, usually metal. Thick is important, because if the detector is too shallow, the shower might leak out the other side, and that means energy that is undetected.

In figure 4.10, we see the effect of adding showering detectors. The depth of a shower depends mostly upon the identity of the particle that is causing it. We note that the neutrino has yet to leave a trace in any detector. Further, the muon

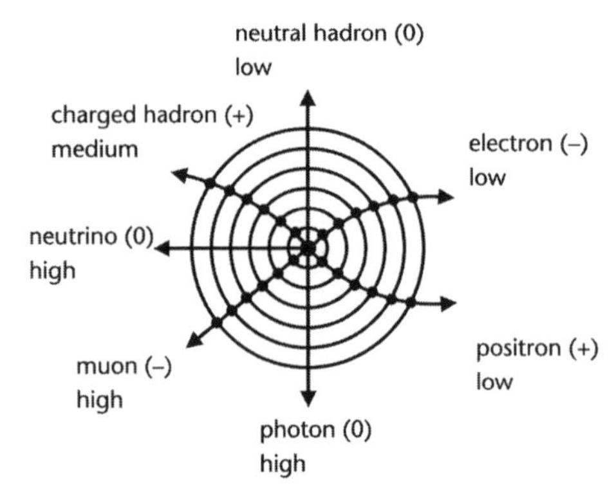

Figure 4.9. The same as figure 4.8, but with a typical ionization detector superimposed. The concentric circles denote six ionization detectors, each in the shape of a cylinder. The dots show where the charged particles traverse the detector and leave a signature. Note that the neutral particles are not observed. The terms low, medium, and high indicate the energy of the particles.

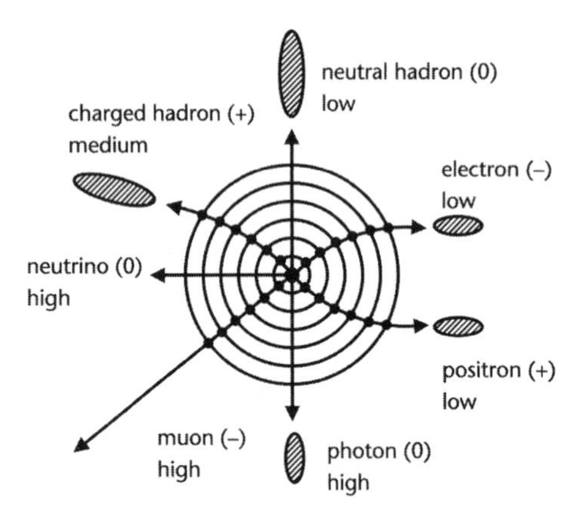

Figure 4.10. The same as figure 4.9, but with the showers added. The ellipses show a shower, with the size of the ellipse showing how deep the particles penetrate. Hadron-initiated showers penetrate more deeply than electromagnetically initiated ones.

didn't shower and passed through the dense material, leaving only ionization energy. Since the magnetic field isn't found in the metal, the muon travels in a straight line there. To make sure the particle is a muon, modern detectors typically have a few additional ionization-based detectors outside the showering detector. There may or may not be a magnetic field where the outer ionization detectors are situated. In our simplified example, let's put a magnetic field there. Figure 4.11 shows this final detector configuration, with the truth knowledge

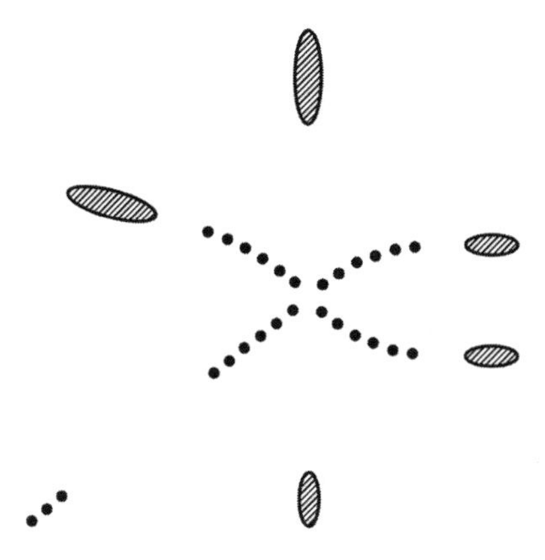

Figure 4.11. The same as figure 4.10, but with all extraneous information removed. Note in the bottom left three additional ionization detector hits, which are from the muon ionization detection system that is outside the shower-measuring device. Compare this figure (which is what is available to the experimenter) with figure 4.7 (which is what we are attempting to observe).

removed. Contrast this to figure 4.7, where the truth information was revealed. Modern particle experimental physicists train to turn what they can see (figure 4.11) into what was there in the beginning (figure 4.7). A real-world example of this is shown in figure 4.26 however, understanding that particular figure requires more information, which we will now turn to. If you're an impatient sort, feel free to peek ahead.

Up until this point, we've been describing detectors in general terms. It is time to get more specific. Our essential questions must be these: (1) How do we measure the position of a particle? and (2) what are the essential components of a showering detector? Let's start with the first question.

Ionization Detectors

The basis for most position-measuring detectors is ionization. Electrically charged particles cross some material, they disturb the electrons in the atoms of the material, and the disturbance is somehow detected. The trick is then to find out where the particle enters the material. While there are a number of choices one can make to accomplish this, by far the most common is to simply make lots of small ionization detectors, isolate them from one another, and just use the information about which ones were hit to indicate where the particle passed.

As an illustrative example, suppose you made a detector somehow composed of ordinary soda straws. Inside each straw is some unspecified material that can be ionized and in which the ionization can be detected. If you took these straws and laid them side by side, they would form the plane seen in figure 4.12.

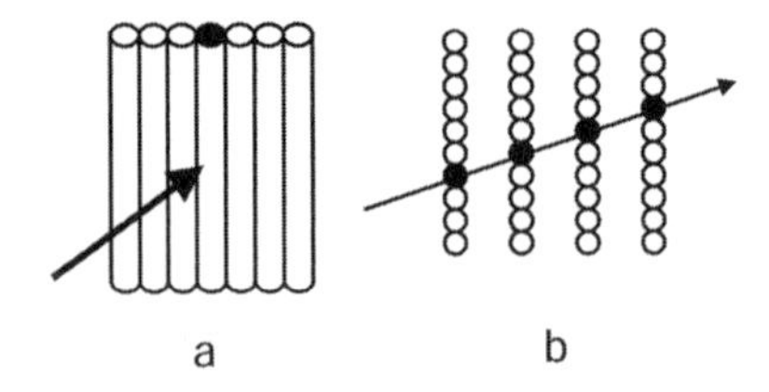

Figure 4.12. Common ionization detectors consist of many smaller detector elements. Individual detector elements are traversed by a charged particle and then signal the particle's passage, with the direction of the particles shown by arrows and the areas they pass through shown by black dots. *a*, face-on view; *b*, a bird's-eye view of multiple planes.

Each straw tells us when it is hit (shown with a black top in the figure). We don't necessarily know where in the straw the particle passes, but at least we know which one and that tells us something about the particle's position. If we have many rows of these straws, we can then measure the particle's path. Figure 4.12a shows a side view of one plane of straws in which it is harder to see which ones were hit. In figure 4.12b, we take a bird's-eye view of many of these planes and see how the pattern of hit straws gives us the information we need about where the particles hit.

Figure 4.12 illustrates an important point. While the whole "straw" technique can give the necessary position information, a big limitation is the size of the individual straws. The bigger the straws, the less precise the position can be measured, while smaller straws give more precise measurements. The reason is simple. Since you know the particle passed through the detector, if it passed through a smaller detector it had to be located in a smaller area. So it is clear that you should make your straws as small as technologically possible, right? The answer is "Yes, but . . ." This important "but" reminds us of the real-world consideration of cost. Each straw needs its own electronic circuit to read it out, and each circuit adds a cost to the budget. More straws means more expense. The reality of fixed budgets imposes a very real limitation on how small one can make the individual straws. In the end, scientists compromise. When precise measurements are critical, they make small detectors and accept the large cost, but when circumstances allow, they use bigger straws and spend their money elsewhere.

In our example, we used a hypothetical detector made of straws. While there are detectors literally made of straws, that technique is relatively rare. More commonly, scientists use two techniques: wires and silicon.

In the former case, the straws are replaced by wires. In fact, a plane of wires looks a lot like a harp. The wires are placed inside a container filled with a carefully chosen gas, and the wire nearest the point where the particle crosses the plane is the one that reports the particle's passage. The space between adjacent wires varies depending on detector design, but a quarter or half an inch (six to twelve millimeters) is reasonable. With sophisticated electronics, detectors using this technology can be made to measure a particle's position with a precision of about a hundredth of an inch or smaller (under 0.3 millimeters).

To measure more precisely, one must turn to silicon technology. In recent decades, engineers have made enormous strides in minimizing the size of electronic chips for use in computers. This technology can be turned to making particle detectors. In silicon detectors, the "straws" are little strips of silicon, a couple of inches (a few centimeters) long and so narrow that you could fit twenty of them in the space of a twentieth of an inch or so (about a millimeter). Recently, it has become economically and technically feasible to make little square "dots" of silicon, 0.002 inches (a twentieth of a millimeter or so) on a side (although supplying enough electronics to read out all these tiny detectors is extraordinarily expensive).

There is a type of detector that actually uses something that really does look like straws. In this case, charged particles traverse literal straws, filled with material that experiences ionization. As charged particles cross the straws, transition radiation is emitted in the form of x-rays. These x-rays then ionize the material inside the straws and allow the particles to be detected. The ATLAS detector uses this technology.

Showering Detector Techniques

With our discussion of ionization detectors complete, we turn our attention to showering detectors. Showering detectors all have one thing in common. They all consist of dense material. Two techniques occur the most often. The first we will discuss is called a *sampling calorimeter*.

A sampling calorimeter's name comes from the fact that it measures a representative sample (sampling) of a particle's energy (calorimeter). The basic structure of such a detector consists of a series of metal plates, separated by material in which it is easy to measure ionization energy. This material can be a gas, solid, or liquid, with a density that is low compared with metal. A typical detector of this form might consist of a plate of steel an inch (25 millimeters) thick, followed by half an inch (12 millimeters) of the material that actually detects. This pattern might repeat fifty times. The details (thickness and materials used) of any particular detector will vary.

Figure 4.13 illustrates the important features of a sampling calorimeter. Particles interact with atoms in the high-density metal and then shower. These

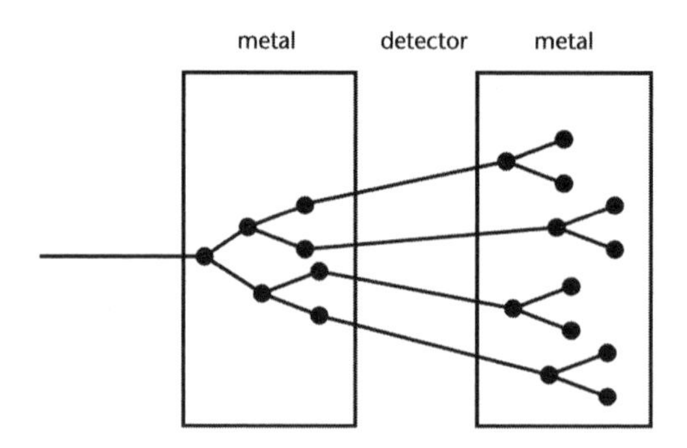

Figure 4.13. In a sampling calorimeter, alternating plates of high density (metal) and low density (gas, liquid, or plastic) are used. The high-density material causes the showers to develop, while ionization energy is measured in the low-density material with the process repeated in subsequent layers.

shower particles then travel through the active detector material, where the ionization is recorded. The showering begins again in the next metal layer. As this pattern repeats itself, the energy of the individual shower particles drops. Eventually, the energy of the particles drops enough that showering stops. These particles are then simply absorbed, and the shower ends. The entire process takes a few billionths of a second.

The second kind of showering detector is called a *homogeneous calorimeter*. The term calorimeter (energy measuring) means the same thing. However, unlike a sampling calorimeter, a homogeneous calorimeter doesn't have any structure. The whole detector is the same. Being homogenous, containing metal, and being able to be read out is quite a trick. This is usually accomplished by using some kind of metal-containing glass. The most common form of this kind of glass is lead crystal. That's right; the same crystal that makes up the chandeliers in a chic hotel's grand ballroom or in that decanter your mom got for her wedding is an ideal material to be used in a particle detector.

A detector made of lead glass works slightly differently from a sampling calorimeter. A high-energy particle enters the glass, traveling faster than the speed of light through the same material. This particle emits Cerenkov light. The particle encounters a lead atom and showers. The daughter particles after the shower are also traveling faster than the speed of light in the glass, so they too emit Cerenkov light. These daughters also encounter lead atoms, and the shower grows. The daughter (and granddaughter and . . .) particles all emit Cerenkov light. Because the detector is made of glass, the Cerenkov light travels to the end of the detector and is collected and converted into electricity. Because of minor details in how showers develop, these kind of metal-containing glass

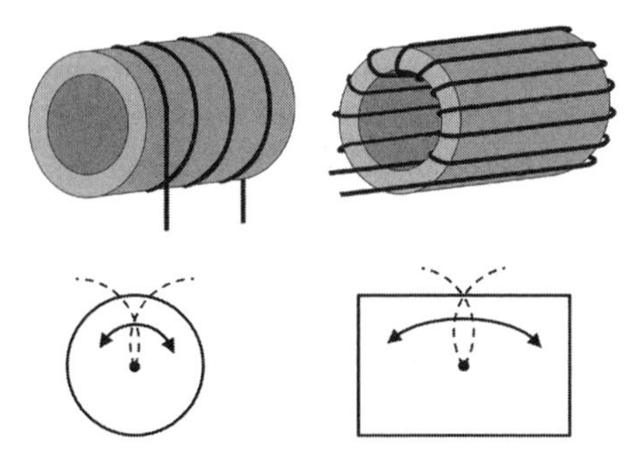

Figure 4.14. Two types of magnets are commonly used in particle detectors: solenoid (*top left*) and toroid (*top right*). Particles will bend depending on the type of magnet used. A solenoidal magnetic field will bend charged particles either clockwise or counterclockwise when the solenoid is seen face-on (*bottom left*), while a toroidal magnetic field will bend the trajectory of charged particles along the side of the magnet (*bottom right*). Figure courtesy of Barry Panas.

particle detectors are predominantly used to detect electrons, antimatter electrons, and photons.

The "Oids:" Solenoids and Toroids

Because two of the LHC experiments have unfamiliar words in their name, we must make a brief detour into how specific magnetic fields can be made. All magnets in modern particle physics experiments are made by wrapping coils of wire in various shapes. Electric current is passed through the coil, and it is it that current that actually generates the magnetic field. These shapes form different kinds of magnetic fields, which in turn bend the particles in different directions. The two most common types of wire-wrapping patterns are *solenoids* and *toroids* (figure 4.14). In the solenoid pattern, the wire is wrapped around the outside of a cylinder in the shape of a spring or a "slinky." A toroid pattern is formed by wrapping the wire in the shape of a bagel. A solenoidal magnetic field bends particles in the plane perpendicular to the beam, while a toroidal magnetic field bends particles toward or away from the beam.

Specific Detectors at the LHC

With our discussion of detector techniques complete, we are now able to outline the specific detectors arrayed around the LHC. The accelerator has two general-purpose detectors designed to study the highest energy proton-proton collisions (CMS and ATLAS), each with associated detectors designed to study much lower energy phenomena (LHC Forward and TOTEM). These detectors can be considered to be "add-ons" to the two main detectors. In addition, two other detectors have been carefully designed to answer much more focused questions: The LHCb experiment is optimized to study collisions in which bottom quarks are produced, and ALICE is designed to study the collisions of heavy ions, for instance when lead nuclei are collided together at nearly the speed of light. Let's introduce each detector in turn.

The rest of this chapter discusses specific detectors located at the LHC. Because we are being specific, there are a lot of details given, for example, sizes and number of pieces that go into the individual detectors. If you're not a detail kind of person, you can skip this description. The principles discussed thus far are featured in various ways in all of the LHC detectors, and so you can have a pretty good, although general, idea how all of these detectors work with what you've learned to this point. For those of you who are detail-oriented, let's plow ahead. We'll meet up again with the "big picture" people at the last section of this chapter, where you can take a peek at some of the tales that insiders swap when they talk about the trials and tribulations of actually putting together these huge detectors.

Two Main Detectors: CMS and ATLAS

Compact Muon Solenoid

The Compact Muon Solenoid, or CMS, detector gets its name because it is relatively small (compact), is optimized to study muons (muon), and has a solenoidal magnetic field at its heart (solenoid). Of course, compact is relative. It's still three stories tall. When we contrast the ATLAS and CMS detectors, the basis of the "compact" term will become more apparent.

Like most modern detectors found attached to particle accelerators, CMS (and ATLAS and ALICE, described below) has a layered "onion" structure, with different types of detectors nestled within one another (figure 4.15). For simple reasons of engineering, the layers in these detectors are roughly cylindrical in shape, looking like nothing more than a large soup can. The various cylinders representing each type of detector technology are nested together like a series of Russian matrioshka dolls.

The CMS detector is 65 feet (20 meters) long and 48 feet (15 meters) in diameter and weighs 14,000 tons. It consists of six distinct layers in the "central" or "barrel" region (i.e., the sides of the soup cans) and five distinct layers in the "endcap" regions (i.e., the top and bottom of the can). In the barrel region, these layers consist of two different types of silicon detectors; a calorimeter to measure the energy of electrons, positrons, and photons; a calorimeter for measuring the energy of the quark-containing "hadronic" particles; a magnet; and finally a system for observing muons. The endcap detectors have the same

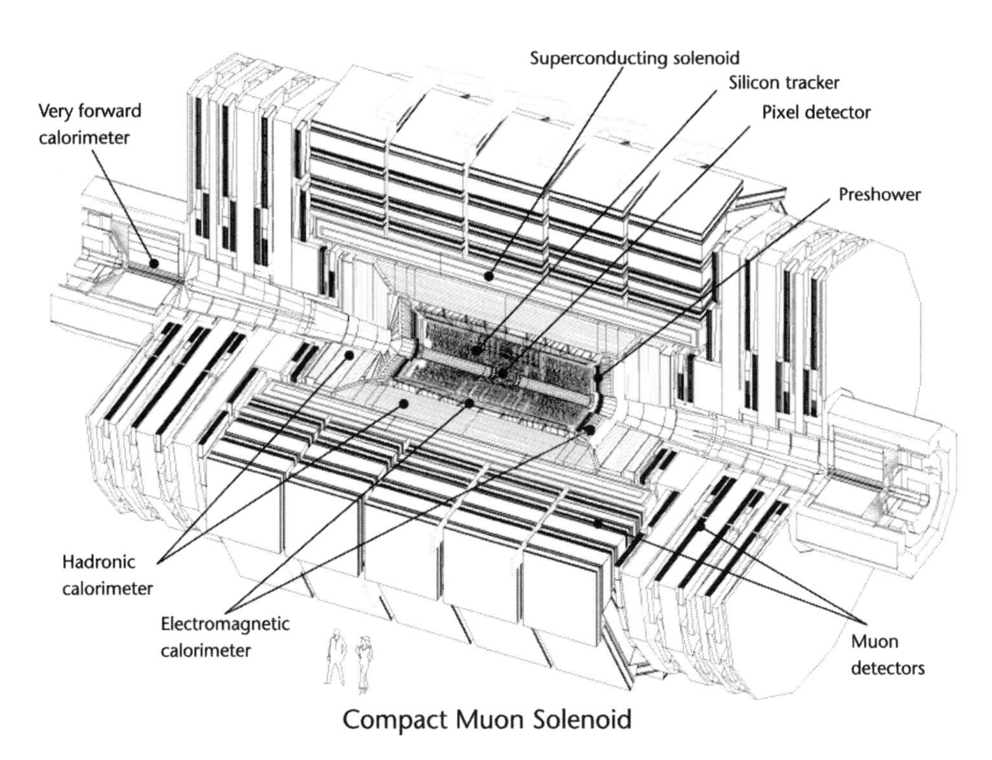

Figure 4.15. A view of the CMS detector, with the subdetectors labeled. Note the size of the people drawn to scale at the bottom. Figure courtesy CERN and the CMS collaboration.

structure except without the magnet. Let's learn something about these various detectors.

The silicon tracking detectors of CMS are simply staggering in their technical parameters and are currently without peer. The silicon tracking detectors sit in a volume about 19 feet (6 meters) long and a little over 6 feet (2 meters) in diameter. This volume is not entirely filled with silicon but rather layers of silicon and air. Each layer of silicon is a tiny fraction of an inch (a fraction of a millimeter) thick and is mounted on a cylinder made of a carbon fiber composite. Note that carbon fiber is used to make modern ultralight airplanes because of its strength. Adjacent cylinders are separated by an inch or two (a couple of centimeters). The support is kept as light and thin as possible to minimize interactions in support material as opposed to material that forms the detector itself.

The silicon system is broken up into two different subsystems, one consisting of tiny silicon detectors and the other of super tiny ones. The inner system is called the *pixel detector*, because it contains super tiny pixels of silicon.

The pixels in the CMS silicon system are incredibly small. In fact if you were to look at the pixel detector under a microscope, you'd see that each square millimeter contained about sixty distinct detectors. With such tiny granularity, the CMS pixel detector consists of about 66 million distinct pixels. This incredible number of detectors is spread over a mere three cylinders, about 36 inches (90 centimeters) long, and ranging in radius from a little less than 2 inches to 4 inches (5 to 10 centimeters).

The second CMS silicon detector is much larger. It is 19 feet (6 meters) long and consists of ten cylinders ranging in radius from about 8 to 40 inches (20 to 100 centimeters). Being so much larger, you'd expect that the number of silicon detectors involved would be much greater, but this apparatus consists of "only" 10 million individual detectors. (I don't know about you, but to use the words "only" and "million" in the same sentence seems weird to me.) The reason that this system contains so few individual detectors is that each detector is much longer. This detector is called the *silicon microstrip detector* because each detector is a thin strip of silicon about 0.006 of an inch (0.15 millimeters) wide, but a couple of inches (several centimeters) long. While in a world without resource limitations we'd have liked to have made just one kind of detector, consisting only of pixels, cost prohibits that option. So, as we learned in our general comments on silicon detectors, you make a finely grained detector when you must and make one with larger individual elements when you can.

The CMS silicon detectors cover an enormous area. In fact, if you took all of the silicon comprising the CMS detectors and laid them edge to edge, they would entirely cover the floor of a 2,200-square-foot (215 square meter) house, with just about enough space leftover so you could stand and enjoy your handiwork.

For the same reasons that your computer needs a fan, the CMS detector needs to be cold to run well. When silicon is warm, it will generate an unacceptable amount of electric current within itself and stop working. Further, we must use electricity to make the silicon detectors work. Like most simple household appliances, the silicon heats up when powered. Thus to work well, the silicon detectors require cooling. When the silicon is working properly it operates at about 10°F (–10°C).

The next detector one encounters as one moves outward from the center is an unorthodox choice. Ordinarily, the next layer would be the coils through which electric current flows to make the magnetic field. However, in CMS the next layer is the calorimeter used to measure the energy of electrons and photons. Since photons and electrons are electromagnetic particles and the device used to measure energy is a calorimeter, this device is called the *electromagnetic calorimeter*, or ECAL (figure 4.16).

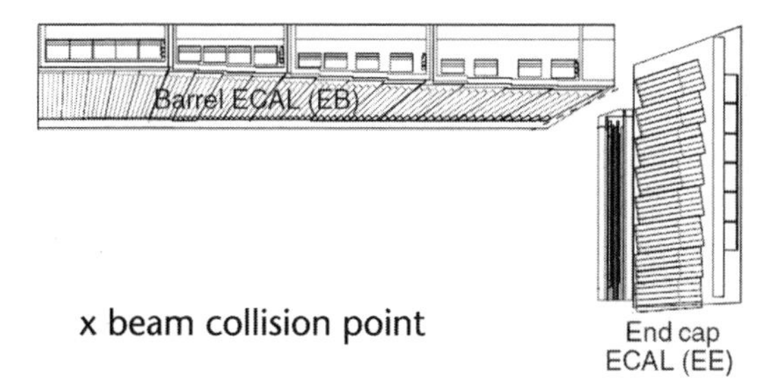

Figure 4.16. This figure shows a side view of the electromagnetic calorimeter, or ECAL. Only a quarter of the detector is shown. The array of rectangles denotes individual blocks made of lead tungstate. Figure courtesy CERN and the CMS collaboration.

The ECAL is an example of a homogenous calorimeter. Rather than the lead glass that was discussed in the overview section, the CMS ECAL is made of blocks of lead tungstate ($PbWO_4$ for the chemically minded). Lead tungstate is amazing stuff. While a casual inspection of one of the blocks used in CMS would lead you to believe that it was ordinary, if rather clear, glass, each block is actually 98% metal by weight. Each block is about an inch (2.5 centimeters) square and about 9 to 10 inches (23 to 25 centimeters) long.

Figure 4.16 shows how these blocks are arrayed around the collision point. The ECAL basic shape is a cylinder, with blocks around the barrel and on the endcaps. The barrel requires 61,200 blocks and the two end walls 14,648 for a grand total of 75,848 blocks. Taken together, the lead tungstate blocks in CMS weigh about 90 tons or about as much as thirteen adult elephants.

The next layer in the CMS detector is the *hadronic calorimeter*, or HCAL. Recall that hadrons are particles containing quarks, of which the proton and neutron are the most familiar, although the type of hadron most commonly produced in particle physics collisions is called a pion. The HCAL is a sampling calorimeter. Like the ECAL, the HCAL is cylindrically shaped with a barrel and endcaps. The metal most used in the HCAL is brass, although steel is used in a couple of places. Recall that a sampling calorimeter requires layers of metal interspersed with layers of material in which the ionization energy is measured. In the HCAL, this low-density material converts the ionization to light, which is converted in turn to electrical signals. Mostly, the light-producing material is a type of plastic—very similar in appearance to Plexiglas. The layers of metal and plastic consist of plates of brass, either 2 or 3 inches (5 to 8 centimeters) thick, followed by about an eighth of an inch (3 millimeters) of plastic.

Recall that the ECAL was made of many blocks of lead tungstate. By knowing which block was hit, you could determine the position where an electron or photon hit the ECAL. The HCAL is conceptually pretty similar, with stacks of

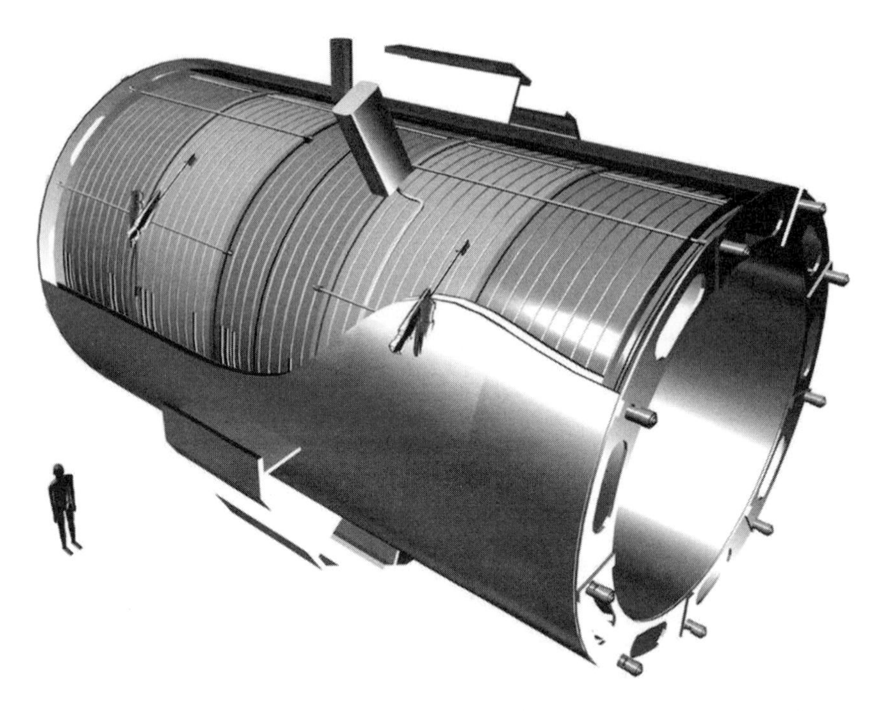

Figure 4.17. The CMS solenoid magnet has the largest physical dimensions of any solenoid ever made and has one of the strongest magnetic fields. Note the size of the person in the lower left of the figure. Figure courtesy CERN and the CMS collaboration.

metal and plastic effectively making blocks. In CMS, there are 2,592 blocks in the barrel of the HCAL and 2,592 blocks on the ends. In addition, there is another calorimeter very near the beam. This calorimeter is made of steel to make the showers and quartz to make the measurement.

Because the creators of CMS made the unusual choice of putting all the calorimeters inside the magnet, the HCAL isn't quite as thick as it should be. This is because to make it thicker, the magnet surrounding it would have needed to be bigger. Since cost considerations and the simple need to transport the detector through city streets to get it to where it was being installed made that choice impossible, a few more layers of calorimeter were added onto the outside of the magnet to catch the "tails" of the hadronic showers. The tail of a hadronic shower consists of the few rare particles that penetrate more deeply than usual. This "add-on" calorimeter is aptly called the *tail catcher.*

Between the HCAL and the tail catcher is the CMS magnet (figure 4.17). The CMS magnet consists of a cylinder with an inner radius of 9.5 feet (3 meters) and an outer radius of 10.5 feet (3.3 meters). The cylinder is about 40 feet (12.5 meters) long and is wrapped 2,168 times with wire. This wire carries the electric current needed to make the magnetic field. The magnetic field in CMS is very strong, about 80,000 times that of the Earth.

In order to make such a huge magnetic field, about 19,500 amperes of current must pass through this wire. In contrast, most houses need fewer than 100 amperes. Thus the CMS magnet alone uses over 200 houses' worth of electricity or about the same as a small suburban neighborhood. To have that much electric current and magnetic field requires an enormous amount of energy (2.7 billion joules for the technically minded) or about enough energy to melt 18 tons of gold. Finally, in order to keep the wire from vaporizing under the onslaught of that much current, the wire of the magnet needs to be made superconducting. Superconducting, as you might recall, is the case where electric current flows without resistance (and thus without heating up the wires.) Making the wire superconducting requires it to be cooled to about –450°F (–270°C).

All these technical requirements pose a serious challenge for the CMS engineers. Let's think a moment about some of the implications of these numbers. With the sheer amount of current that is required, special accommodations must be made for the power, with a special substation whose sole purpose is to power the CMS site. In addition, while the wires of the magnet must be –450°F (–270°C), the outside of the magnet must be at room temperature. This means that the magnet must essentially be a large Thermos bottle, different only in size from the one that keeps coffee hot.

Another engineering consequence of the design of the CMS magnet has to do with an inherently self-destructive aspect of designing a large electromagnet. Current makes the magnetic field. However current in a magnetic field feels a force. That's how electric motors work. So here we have wires that carry current that make a magnetic field. They in turn feel a force and thus want to move. The force is about 2 to 3 tons for every foot-long (30 centimeter) segment of the wires that make up the coils. Thus in order to make the highly precise magnetic field, each foot of wire must withstand the force equivalent to the weight of two or three mid-sized American cars. Recall that the coils comprise 25 *miles* (40 kilometers) of wire and you get an idea of the kinds of distorting forces present in the CMS magnet.

Outside the magnet is the tail catcher briefly mentioned before. But the big system outside the magnet is the series of muon detectors. Because all particles except muons are stopped by the calorimeters, the environment in the muon systems is relatively benign. Unlike the detectors closer to the beam, into which thousands of particles plow every 25 billionths of a second, the muon detectors see only a few.

But for all that, the muon system is still challenging. By far, the muon system

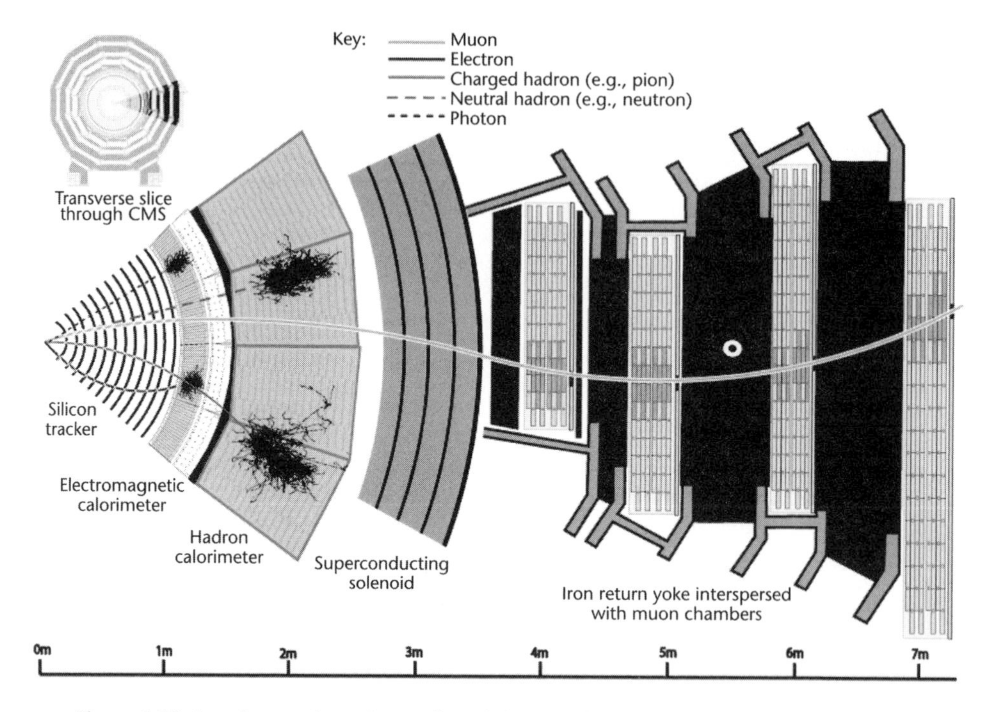

Figure 4.18. An edge-on view of a portion of the CMS detector depicting all important detector components and the response of the detector to various types of particles, with its location on the detector shown in the inset in the upper left. The muon's trajectory curves clockwise to the left of the solenoid and counterclockwise after the solenoid, because the direction of the magnetic field has changed. The iron return yoke cycles the magnetic field back through the detector. Figure courtesy CERN and the CMS collaboration.

is the largest of all of the subdetectors. In the barrel region, it ranges from a radius of about 12 to 24 feet (3.75 to 7.5 meters) and is about 21 feet (6.6 meters) long. The endcaps range from 18 to 34 feet (5.6 to 10 meters) long and have radiuses from about 4.5 to 22 feet (1.4 to 6.9 meters).

The muon system consists of four thick slabs of iron, interspersed by four layers of position-measuring ionization detectors. Each of these four layers actually consists of many smaller layers. When all of the individual detectors in the muon systems are counted, they number about 830,000. Figure 4.18 shows a slice of the CMS system, drawn to scale.

Like everything at the LHC, the CMS detector is not static. Scientists are constantly planning how to improve it. The reasons for the need to change are many, for instance, lack of enough money to build the full detector in the beginning, improvements in available off-the-shelf technology, and evolving beam conditions all contribute to the need for making these changes.

The detector described here is the configuration in its first active phase. During the first long shutdown (LS1, spring 2013 to fall 2014/spring 2015), the detector will get another plane of muon detectors in the endcaps (the top and bottom of the soup can), and the electronics that convert light to electricity will be changed in the hadron calorimeter, both in the tail catcher and in the endcap calorimeters. A smaller-diameter beam pipe will be installed, in preparation for a new pixel detector the following year.

During the brief CERN shutdown in Christmas 2016, the pixel detectors will be replaced. Rather than the current three barrels (the sides of the soup can) and three layers in the endcap, there will instead be four barrels and four layers in the endcap.

During the second long shutdown (LS2, 2018 to 2020), the hadron calorimeter electronics will be changed, including the equipment that converts light to electricity. During this shutdown, the electronics that decide which events should be recorded will be replaced.

In the third long shutdown (LS3, 2022 to 2025), the entire tracker will be replaced. After years of operation, the existing detector will have undergone enough radiation damage to no longer function well. While many of the necessary specifications have been already worked out, as of this writing, the final design is still incomplete.

ATLAS

The second big detector we will discuss at the LHC is ATLAS (A Toroidal LHC ApparatuS; figure 4.19). The intent of this detector is to study the same kind of collisions as CMS, with different design choices. Given that nobody knew what kinds of new physical phenomena would be found at the LHC, it seemed prudent to have two competing detectors using different technology. In addition, a second detector allows independent verification of any discovery and provides some insurance in case some catastrophe (like a fire or flood) destroys a detector. And, of course, nothing motivates like competition. Two detectors keep the researchers from becoming complacent.

While the two detectors have appreciable differences, they also have some broad similarities. This isn't that surprising, since both detectors are designed to do the same thing; namely sit at a spot where two beams of protons intersect at their heart and sift through the twenty to a hundred collisions that occur every 25 billionths of a second and look for something never before seen. So broadly, both detectors are large cylinders, with barrels and ends. At the heart of both CMS and ATLAS are silicon detectors, sitting in a magnetic field. This silicon is surrounded by the energy-measuring calorimeters, followed by muon detectors.

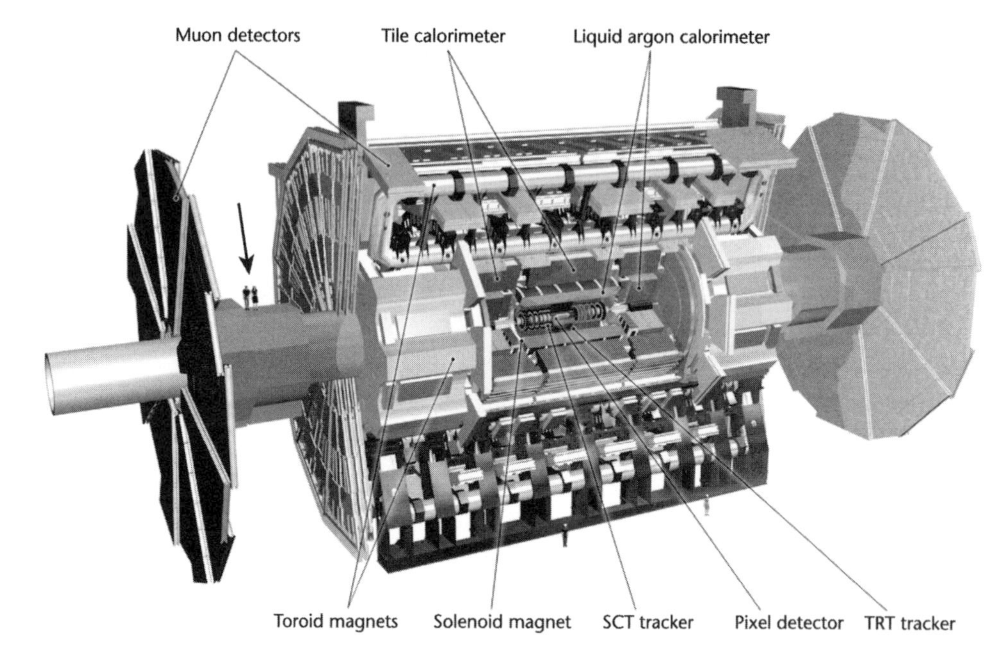

Figure 4.19. The ATLAS detector with components identified. Note the tiny human images on the left between the two muon detectors and two more at the base of the detector. SCT = Semiconductor Tracker; TRT = Transition Radiation Tracker. Figure courtesy CERN and the ATLAS collaboration.

However, for all their similarities, the two detectors are quite different in detail. The first difference is the physical size. ATLAS is much larger than CMS. While CMS is 65 feet (20 meters) long and 48 feet (15 meters) wide, ATLAS is 150 feet (47 meters) long and 75 feet (23 meters) wide. ATLAS's volume is about six times larger than CMS. In spite of its size, ATLAS is also much lighter, with a weight of about 7,800 tons (compared with CMS's 14,000 tons).

ATLAS's much larger size stems from the designers' choice to focus on muon measurements. ATLAS's muon detectors can operate alone, while CMS requires both the muon detectors and the silicon tracker to measure the properties of muons created in the particle collisions. And, as they say, time will tell which choice was best.

The center of the ATLAS detector also consists of silicon pixels. These pixels are about 0.002 × 0.016 inches (0.05 × 0.4 millimeters) in size. The pixel detector consists of three layers, spread out in a cylinder ranging from about 2 to 10 inches (5 to 25 centimeters) radially and about 4 feet (1.3 meters) long. The ATLAS pixel detector consists of 80 million pixels, somewhat more than CMS's 66 million.

Outside the volume filled with the pixel detector, the ATLAS group has chosen to put another silicon-based detector. Like CMS, the size of the individual silicon detectors is much larger in this region. These silicon strips are 0.003 inches wide but about 5 inches long (0.08 × 130 millimeters). This detector consists of about 6 million individual silicon detectors. This detector sits in a cylindrical volume ranging from a radius of 1 to 2 feet (30 to 60 centimeters) and about 18 feet (5.6 meters) long.

Up to this point, the ATLAS and CMS detectors are broadly similar. However, while the CMS detector contains another silicon-based detector filling out a cylindrical volume with a 40-inch (one meter) radius, the ATLAS group chose to use a different technology to fill in this volume.

The next technology encountered as we travel outward from the center of the ATLAS detector is the *transition radiation detector*. The transition radiation detector fills the cylindrical volume over a radius from 24 to 41 inches (61 to 104 centimeters) and is about 18 feet (5.6 meters) long. Its basic construction consists essentially of long straws, 0.16 of an inch (4 millimeters) wide and about 28 inches (71 centimeters) long. Eight of these straws placed end-to-end cover the entire length of the volume, and filling the entire volume requires 350,000 straws.

These straws are filled with a gas mixture that is mostly xenon. As charged particles cross the straws, they ionize the gas and are detected. However, for very fast particles (usually electrons or their antimatter partners) x-ray transition radiation is also emitted. This x-radiation also ionizes the xenon gas, leaving an even bigger signal. By seeing which straws are hit, one can follow the trajectory of charged particles through the volume, and, by seeing which ones have higher or lower signals, you can determine which trajectories are caused by electrons.

While the CMS group made the unusual choice to follow the tracker with the lead tungstate ECAL, the ATLAS group made a more traditional choice. The next layer in the ATLAS detector is the central (solenoidal) magnet.

The central magnet in ATLAS is a solenoid style similar to that of CMS. The magnet fills the cylindrical volume with a radius a little less than 4 feet (1.3 meters) and is about 4 inches (10 centimeters) thick. The magnet is also 17 feet (5.3 meters) long. The wires used to carry the current to energize the electromagnet wrap 1,173 times around the outside of the cylinder and carry a little under 8,000 amperes of current. The net result is a magnetic field of about 40,000 times the Earth's magnetic field, or half the magnetic field at the heart of the CMS detector.

Following the ATLAS central magnet are the two calorimeters, the electromagnetic and hadronic. Both calorimeters consist of the usual barrel and endcap geometry. The electromagnetic calorimeter is of the sampling style and uses lead to create the showers and argon, chilled to a liquid form, to measure the shower energy. The ATLAS electromagnetic calorimeter consists of about 175,000 individual detectors, more than double that of CMS.

The ATLAS calorimeters used to measure hadrons are also sampling calorimeters. In the barrels, layers of iron and ionization-detecting plastic are interleaved, while in the endcaps, the structure is copper interleaved with liquid argon. About 19,600 individual detectors make up the ATLAS hadronic calorimeters.

It is when we turn our attention to the final layers of the ATLAS detector, the muon detectors, that we see the greatest contrast with CMS. To begin with, we finally encounter the large toroid magnets that are featured so prominently in ATLAS's name. The outer ATLAS magnets are enormous. In the central barrel region, the magnets cover a cylindrical volume 80 feet (26.6 meters) long and a radial distance from 15 to 32 feet (4.7 to 10 meters). The endcap toroids are 16 feet (5 meters) long and fill the radial volume from 2.5 to 18 feet (0.8 to 5.6 meters). In both sets of magnets, the current is 20,500 amps. These are big magnets.

The ATLAS muon detection system is similarly impressively large. Various detector technologies record the ionization caused by the muon's passage. About 1.1 million individual detectors comprise the ATLAS muon system. Figure 4.20 shows a slice of the ATLAS detector.

The ATLAS collaboration is also undertaking a vibrant program of detector upgrades and modifications and planning for more. As with the CMS description, the detector described here is the configuration when the book went to press. During the first long shutdown (LS1, spring 2013 to fall 2014/spring 2015), the detector will get another plane of muon detectors in the endcaps (the top and bottom of the soup can). A smaller-diameter beam pipe will be installed and a fourth layer of silicon pixels will be installed.

During the second long shutdown (LS2, 2018 to 2020), new muon detectors in the endcaps will be installed. These will give more information to the electronics that decide which events should be recorded. In fact, there is an extensive program planned to revamp considerable swaths of the decision-making electronics. Either during or before LS2, additional detectors will be placed at 672 feet (210 meters) along the beam pipe from the collision point to study collisions in which one or both of the protons are not hit hard enough to break apart.

In the third long shutdown (LS3, 2022 to 2025), the entire tracker will be replaced just as is the case with CMS. While many of the necessary specifications have been already worked out, as of this writing, the final design is still incomplete. There are plans for considerable upgrades to calorimeter electronics, improvements to infrastructure, replacement of the calorimeter nearest the beam

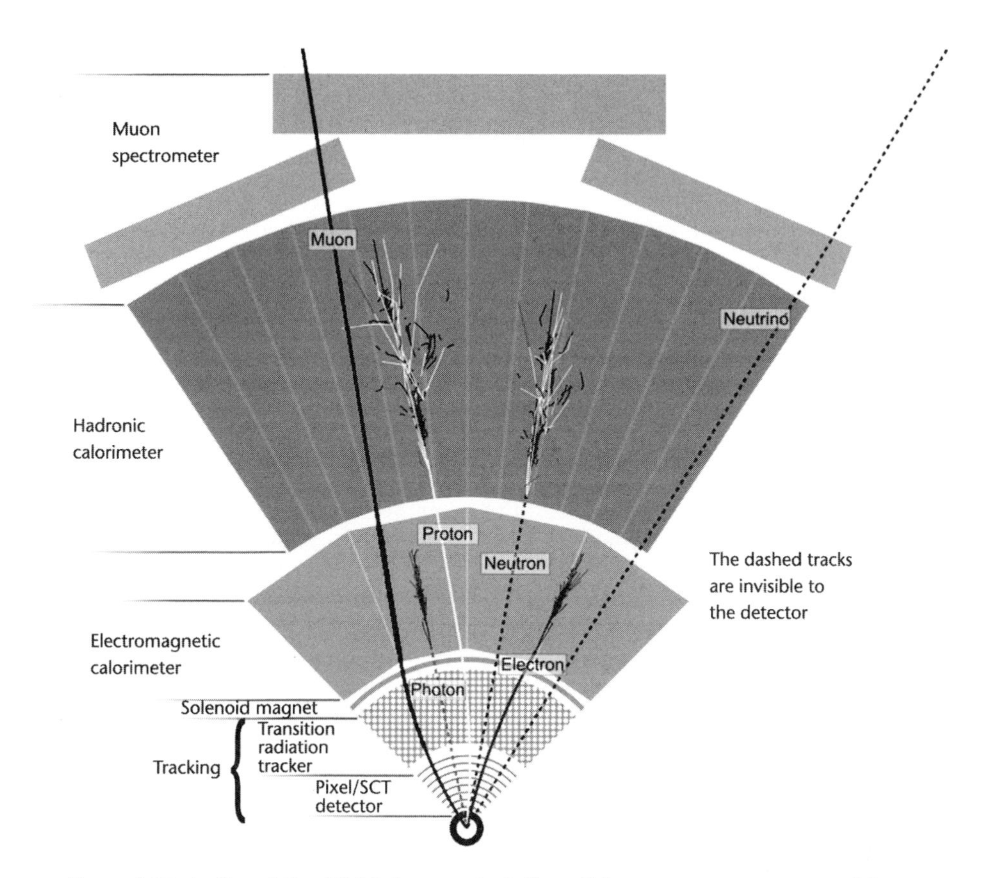

Figure 4.20. A slice of the ATLAS detector, including all important components and how they respond to various particle types. SCT = Semiconductor Tracker. Figure courtesy CERN and the ATLAS collaboration.

and electronic changes to the muon system. Many details aren't yet known and will no doubt change.

ATLAS and CMS: Summary

The two large general purpose detectors at the LHC are amazing feats of both engineering and technology. Both detectors are designed to search for new physical phenomena hidden in the deluge of more pedestrian collisions between protons. Table 4.1 above summarizes the main points of each detector, each containing nearly 100 million distinct detector elements. Only time will tell if one group has made better design choices than the other. If history is any guide, both detectors will have an edge over the other in some particular measurements and yet both will make competitive (and superb!) measurements. We already know that these claims are true, but the real test will occur when an unexpected phenomenon is encountered.

Triggers

One thing we've not discussed is the rate at which the two large detectors can collect data. Recall that the proton beams are set up to allow collisions to occur in the center of both detectors every 25 billionth of a second. That means collisions occur 40 million times each second. Further the beams are intense enough that we expect twenty or so collisions every time they pass through one another. That means that there are about 800 million collisions per second in each of the ATLAS and CMS detectors.

It turns out that each experiment can record and process about 300 events per second, a far cry from the 40 million. Roughly speaking, each detector can record only one collision out of every hundred thousand. The limitation here is cost. While more collisions can be recorded, it's just prohibitively expensive to buy enough computers to do the subsequent analysis.

There is another feature one must consider when attempting to record the interesting collisions: they are extremely rare. Most collisions at the LHC will be relatively gentle impacts between two protons passing by one another, like two strangers brushing shoulders as they pass one another on a street in New York City. However, like the beginning of many a light romantic comedy, in which two hurrying people run head on into one another, occasionally two protons collide violently and some noteworthy physical process is revealed.

The problem is that at the LHC the gentle collisions are about 100 trillion times more likely than the rare ones are. Combined with the fact that the LHC experiments can record and analyze only one event out of a hundred thousand, that means that one has to be very careful in selecting just what collisions to record. The process whereby one selects events is called a *trigger* and is crucial to running a successful experiment. If you choose the wrong collisions to record and you don't have the right data to analyze, you might as well pack it up and go home.

Triggers in a particle physics context are highly complex and fluid, so it is impossible to describe them in detail here. The two experiments have made different choices shaped by the strength and weaknesses of their detectors, and it is a certainty that by the time you read this, the triggers will be different in detail from those the groups have employed as I write this. However, some essentials will remain.

The essence of triggering is having a multiple level scheme. Experiments have two to four levels. The basic idea is that the data flow into electronics (either custom-built or off-the-shelf computer components). These electronics are programmed to evaluate the data, decide whether they are interesting, and pass them along for recording the subset of data that seems like it might be worth keeping. Each level makes ever-more-sophisticated decisions.

As an example, we can think what a two-level trigger might be. Level 1 might look to see if detectors in the muon-detection system were hit by the passage of a charged particle. If you're studying physical phenomena that produce a muon, well then, you can immediately exclude recording collisions in which the muon detectors are silent. The level 1 trigger will make this decision and pass on the events that qualify to level 2.

If it turns out that the muon systems have indicated the passage of a charged particle, this doesn't necessarily mean that you want to record the event. Recall that the fraction of events that you can record is very small. So the level 2 trigger will look at the subset of events that passed level 1 and stare at them a little harder. Since the number of events entering level 2 is now relatively low, it can spend more time and determine the muon direction, energy, and whatnot. Then level 2 will decide whether the muon passes your criteria and will tell the electronics to discard the event or to record it.

The actual trigger system is much more complicated and looks at different facets of the collisions. But the important points are the same as discussed above. To be clear on the language, an event is when the two beams cross and interact. In each beam crossing, twenty to one hundred pairs of protons collide, with generally only one collision being interesting. Thus an event usually includes many collisions, with typically only one collision leaving striking features in the detectors. When professionals say "event," they mean the "interesting collision" and keep in mind the simultaneous collisions that you the reader can ignore. But the pros can't. Roughly speaking, the 40 million beam crossings per second are presented to level 1 for consideration, and it selects about 100,000 events per second as being potentially interesting. Level 2 looks more closely at these 100,000 events and chooses 100 to 300 to record to tape. Later, scientists study these collisions in detail, hoping to see something worthy of note.

Thus we see that the trigger system is a crucial piece of the design of an experiment. Just like a poor choice in the energy or particle types in your accelerator or a poor technology choice in your detector can make your experiment a failure, so too can a poor trigger choice. The number of right choices one must make to simply record the data is rather daunting.

The Special Purpose Guys

While the ATLAS and CMS detectors are the large multipurpose detectors that were the primary reason the LHC was built, there are other experiments at the

LHC, two of which we'll mention only in passing and two of which we'll discuss in a little more detail.

While the experiments are attempting to study the rare collisions that may signal new physical phenomena, these particular collisions do not make up the bulk of collisions that occur. Recall that the "interesting and rare" occurrences are about a part per 100 trillion of the collisions. Some scientists are more focused on the common. After all, nobody has measured these common processes at these energies before.

One of the most common things that can happen when protons collide is they act like two billiard balls, just bumping into each other. Two protons enter the collision and two exit. Because these collisions are relatively gentle, the protons are not scattered at large angles and so don't hit the ATLAS and CMS detectors.

Thus two "add-on" experiments were built that piggyback on ATLAS and CMS. These experiments are small ionization detectors, located near, but outside of, the two big detectors. They record the passage of protons gently bumped out of the beam pipe. The TOTEM detector is associated with CMS, while the equivalent for ATLAS doesn't yet have a snazzy name. In addition, associated with the ATLAS detector is the LHC forward, or LHCf, detector. This detector is located about 550 feet (172 meters) from ATLAS and is designed to look at neutral particles generated very near the beam. These data will help us understand the common things that happen when protons slam into one another and increase our knowledge of the cosmic ray collisions discussed in chapter 7. These detectors are very small (a few cubic feet or a fraction of a cubic meter) and are situated tens to hundreds of feet (several meters to about a hundred meters) from the big detectors and oriented on the beamline.

Two other major detectors at the LHC remain. While the ATLAS and CMS detectors are designed to be general purpose and versatile, general purpose usually means compromise. Being able to do everything usually means that you don't do any particular thing as well as you would if you focused on it exclusively.

LHCb

The ATLAS and CMS detectors are designed to run in the punishing collision environment of having twenty to one hundred or so proton-proton pairs collide at the same time. These detectors must sift through the debris, looking for some rare physical phenomena. Further, this process repeats itself 40 million times a second. The reason one would design an experiment to run under these conditions is that new physical phenomena are very rare, say one interesting collision for every 100 trillion boring ones. Consequently, in order to have a prayer of seeing anything new and exciting, you need to simply collide as many proton pairs as possible and hope for good luck. Further, if you want to make sure you understand the new phenomena you see, you need to wrap your detector like a sphere (or a cylinder in the case of ATLAS and CMS) around the entire collision point.

However, for different physics studies, you'd make different choices. Nowhere at the LHC is this point made more apparent than in the Large Hadron Collider Beauty, or LHCb, experiment. Its main purpose is to study particles containing bottom quarks. These particles are called *b-hadrons*, where "hadron" means "particles containing quarks" and "b" reminds us that at least one of the quarks is a bottom quark. None of these b-hadrons are something you've likely to have heard of before, because they live only briefly . . . usually about a trillionth of a second.

And yet these short-lived particles can reveal fascinating clues about the universe. Studying a class of b-hadrons containing a quark and antiquark, one of them of the bottom quark type, is beginning to shed light on the question of why the universe seems to be composed essentially entirely of matter, a topic we return to in chapter 7. Further, it is thought that by precisely measuring the production and decay of b-hadrons that scientists might discover the Holy Grail of something new.

Because events in which b-hadrons are produced are relatively common (occurring about 1% of the time), we don't have to collide nearly as many protons together to study these kinds of collisions. Recall that, during normal ATLAS and CMS running, about twenty to a hundred proton-proton collisions occur simultaneously. To accomplish this frantic rate, physicists must make very intense beams and put many protons in each.

However, in order to study the production of b-hadrons, physicists still use proton beams but ones that are much less intense than those needed by ATLAS or CMS. In these beams, usually only one pair of protons collides at a time. The reduction in intensity is accomplished by using magnets to make the beam wider as it passes through LHCb (and compressing it when it exits). While the proton collisions still occur 40 million times a second, one collision at a time is much simpler than twenty. In addition, the fundamental philosophy of the LHCb experiment is very different than ATLAS or CMS. ATLAS and CMS want to record and inspect the entire collision. If some new physical phenomenon is observed, the best way to understand it is to record all of the debris from the collision and inspect it all.

The LHCb experiment cares mostly about studying b-hadrons. As long as the collision makes a b-hadron or two, that's a collision its researchers might like to study. A corollary of this choice is that the LHCb experiment doesn't care about recording all the debris from a collision. As long as the b-hadrons are recorded and measured accurately, that's good enough. Further, for technical

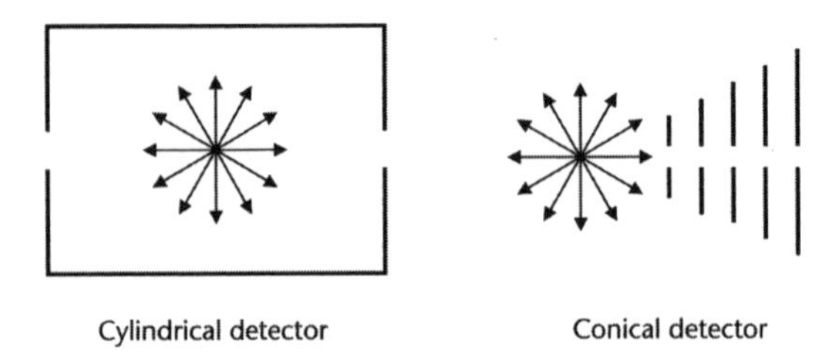

Figure 4.21. Two distinct detector geometries are common in particle physics experiments. A cylindrical detector envelopes the debris of the collision, while a conical detector only samples the debris. These techniques each have their merits, depending on the needs of the measurement.

reasons beyond the scope of this book, when b-hadrons are produced, they tend to be "near" the beam. "Near" means they tend to be produced in a narrow cone of about 40°, oriented on the beamline.

Consequently, the LHCb experiment has a much different geometry from that of ATLAS or CMS. Rather than a cylinder that envelopes the collision point, LHCb is cone-shaped and oriented to one side of the collision point (figure 4.21). B-hadrons fly into the LHCb detector and are analyzed. Many particles created at the LHCb collision point entirely miss the detector. That's OK, as long as the b-hadrons are recorded. With these introductory remarks in mind, we are now ready to look at LHCb in a little more detail.

The LHCb detector is shown in figure 4.22. You'll note its similarity to figures 4.18 and 4.20. LHCb looks like a slice of its larger counterparts.

Nearest the collision point is the Vertex Locator, or VELO, detector (figure 4.23). The name stems from the fact that the detector was designed to observe and measure the vertex caused by the b-hadron decay. The VELO detector is made of silicon, with strips ranging from 0.002 to 0.005 inches (0.04 to 0.10 millimeters) wide. Altogether, 172,000 strips of silicon make up this detector, compared with the 60 to 80 million detector elements in ATLAS and CMS. The VELO detector consists of twenty-one distinct layers of silicon, circular in shape and with a hole down the center. The detector has a passing resemblance to a series of Blu-ray disks stacked and spread out over 3.2 feet (1 meter) or so.

The designers of the VELO system made one interesting engineering choice. Unless great care is taken, silicon is susceptible to damage by radiation. In normal operation, the beams would pass through the center of the hole in the center of the VELO disks. However, when the beam is being put into the LHC, it is larger and the danger of steering it incorrectly is greater. Thus the VELO is split into a left and right side, and the two sides can be retracted during the dangerous

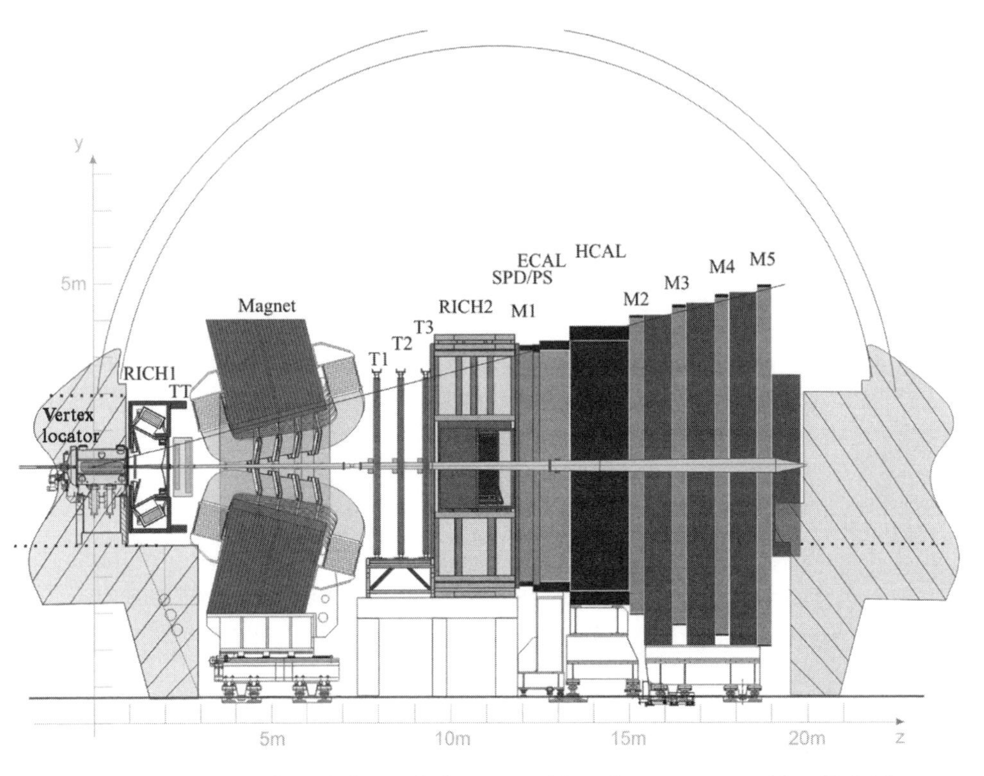

Figure 4.22. An overview of the LHCb detector, with major components identified. T1–T3 are the tracking chambers, M1–M5 are the muon detectors, and SPD/PS is a part of the energy-measuring calorimeter system. ECAL = electromagnetic calorimeter, HCAL = hadronic calorimeter, RICH = Ring Imaging Cerenkov detector; TT trigger tracker. Figure courtesy CERN and the LHCb collaboration.

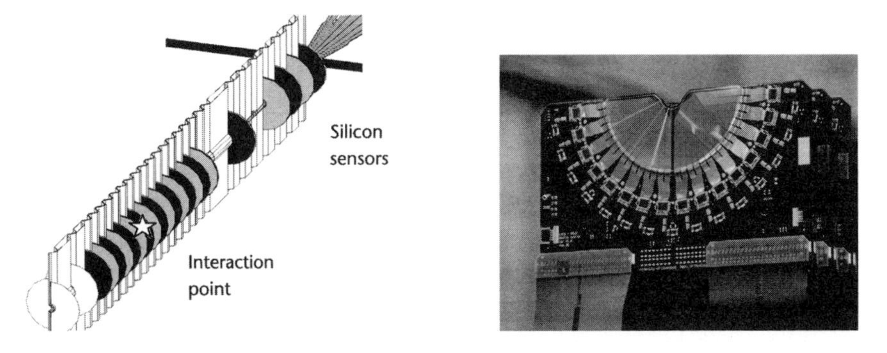

Figure 4.23. The LHCb VELO is a silicon-based detector system that sits extremely close to the LHC beam. Twenty-one disks of silicon surround the interaction point. Figure courtesy of CERN and the LHCb collaboration.

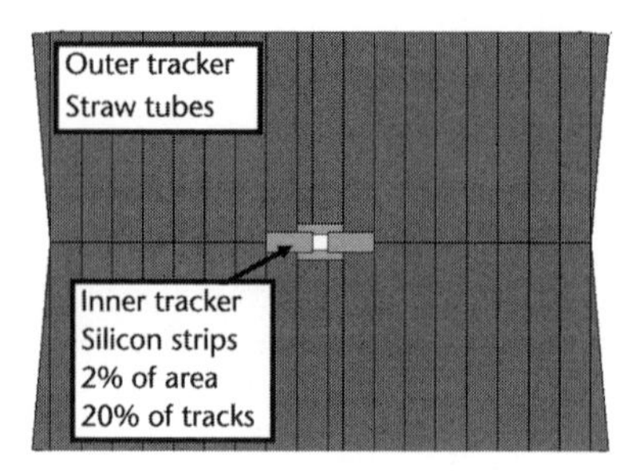

Figure 4.24. The LHCb tracker differs in geometry from those of the other three detectors. This tracker consists of a plane of detectors, consisting of two technologies, with a small silicon detector at the heart, surrounded by a larger tracker consisting of straw tubes. Figure courtesy of CERN and the LHCb collaboration.

moments. In normal running, the VELO detector is positioned a scant 0.3 inch (8 millimeters) from the beam.

The next detector the debris encounters is designed to help identify precisely which particles are passing through the LHCb detector. B-hadrons can decay into this particle or that, and precisely measuring how often the various possible decays occur is one of LHCb's goals. This detector is called a Ring Imaging CHerenkov detector (they used the more phonetic version of Cerenkov for their acronym), or RICH-1. The "1" is because there is a second RICH in LHCb (obviously RICH-2). The rest of the name comes from the fact that Cerenkov light comes out in the shape of a cone surrounding the particle's passage through the material. Depending on how fast the particles cross the detector, the cone will be bigger or smaller. This cone of light hits detectors that convert the light to electricity and leaves a circular pattern. So, by measuring the energy of the particle and the size of the circle, one can frequently identify precisely what particle it is. It takes 200,000 photon detector elements in RICH-1 to properly reconstruct the circular patterns.

RICH-1 is followed by the trigger tracker (TT). This device is made of 180,000 strips of silicon arrayed in four layers. The layers are about 4 feet (1.3 meters) high and about 5 feet (1.6 meters) wide.

The TT is followed by a strong magnet whose job is to bend the path of charged particles traversing it. Once these paths are bent, the charged particles traverse the main tracking system. This system consists of twelve planes, grouped into three stations, each 15 feet (4.7 meters) high and 19 feet (5.9 meters) wide. This tracking system, depicted in figure 4.24, consists of the inner and outer trackers (IT and OT). The inner tracker covers only 2% of the total area near the beam. However this 2% of the area, while small, is where the particles are most concentrated and captures a full 20% of the particles coming out of the collision. The inner tracker is made of 129,000 silicon strips, about 0.008 inches (0.2 millimeters) wide and 4 or 8 inches (10 or 20 centimeters) long. The outer tracker is made of long tubes, filled with a gas that ionizes when charged particles cross them. The outer tracker covers the bulk of the area (98%) with 54,000 tubes.

The tracking system is followed by the second RICH (RICH-2). Its purpose is similar to that of RICH-1: to help precisely determine the identity of the particles crossing it. RICH-2 contains 295,000 detector elements.

Just like the other big detectors, the LHCb tracking is followed by the calorimeters and muon-measuring systems. In most detectors, they occur in that order. But in LHCb, the calorimeters and muon system are intermixed, with the first layer of the muon system coming before the calorimeters.

The LHCb calorimeters are pretty traditional and are separated into an electromagnetic and a hadronic part, both of the sampling variant. The electromagnetic calorimeter is made of lead to make the showers and plastic to measure the ionization. The hadronic calorimeter consists of layers of iron and plastic. Taken together, the calorimeters consist of a relatively modest 20,000 detector elements.

The final LHCb detector is the five-layered muon system that straddles the calorimeter, with one plane coming before the calorimeters and four after. Mostly, the muon system consists of large planes of wires looking like an enormous harp. The wires of the harp are surrounded by a special gas that is ionized when crossed by charged particles. The wires carry the ionization energy out to waiting electronics. A small portion of the first layer of muon detectors consists of a technology that allows a more precise determination of the position of the muon's passage. The muon system's five planes each consist of about 25,000 different wires to read out muon position.

The fact that the LHCb experiment contains something like 1% of the pieces of ATLAS or CMS could be taken to mean that the LHCb experiment is inferior. Nothing can be further from the truth. For about 15% the cost of either of the other two larger detectors, LHCb will capture more than double the number of b-hadrons in any one collision compared with its larger brethren. This is because the LHCb detector was designed especially to study b-hadrons. When you are not designing for all measurements (and therefore having to be all things to all people), you can focus on your core game and perform better in your own little niche compared with your broader-scoped neighbors.

The LHCb detector will be upgraded over the future, with the bulk of the upgrades being done during the second long shutdown (LS2, 2018 to 2020). The trigger will have to be upgraded. If not, the electronic rates will be saturated. Saturated electronics means that they might not be active when a rare physics event occurs. It's kind of like a juggler with the maximum number of balls in the air. If you try to toss him another ball, he just won't be able to grab it. Upgrading the trigger will require new electronics reading out all of the detector subsystems except for calorimeters and muons.

The RICH detectors will be replaced, as will the VELO system and the other silicon-based trackers. The system in RICH-1 will be replaced with a new detector called TORCH, which uses Cerenkov light and specifically looks at the time at which the light arrives. The light from slow-moving (and thus heavy) particles will arrive after the lighter particles. TORCH might be installed after the other upgrades.

The replacement of the VELO system will have to be resistant to radiation damage, as the detectors are very close to the beam, and it is a punishing environment. Further, the higher beam luminosity means many more collisions, which means more particles crossing the detector. Accordingly, the VELO replacement will require smaller individual pixels to better identify tracks that are near one another. The TT system and parts of the outer tracker system will be replaced probably by silicon, although as of this writing the technology has not been selected.

ALICE

The last of the big detectors at the LHC that we discuss in detail is the ALICE (A Large Ion Collider Experiment) detector, shown in cross section in figure 4.25. Unlike the other three, which are optimized to study the debris of the collisions of proton pairs, ALICE is optimized to study collisions involving lead nuclei. Lead nuclei consist of 208 protons and neutrons, and thus these collisions are much more complex. The energy of the lead beams is immense. While for ordinary proton-proton collisions, the LHC can collide particles with an energy of 14 trillion electron volts of energy, in the case of lead collisions, the total energy is more like a quadrillion electron volts.

Given that the total energy involved in lead-lead collisions is 82 times larger than the proton-proton collisions, you might be wondering why people even bother with ATLAS and CMS. The reasons are straightforward. First, the energy in lead collisions is shared between many protons and neutrons. Thus the energy isn't as concentrated in lead collisions. It's like comparing the amount of heat in a room with that of a flaring match. There's much more heat in the room, but the match can still burn you.

The second reason is that it is difficult to make intense beams of lead. Even working rather hard, the lead beams will be about 10 million times less intense than the proton beam case. These beams are so diffuse that, even though they

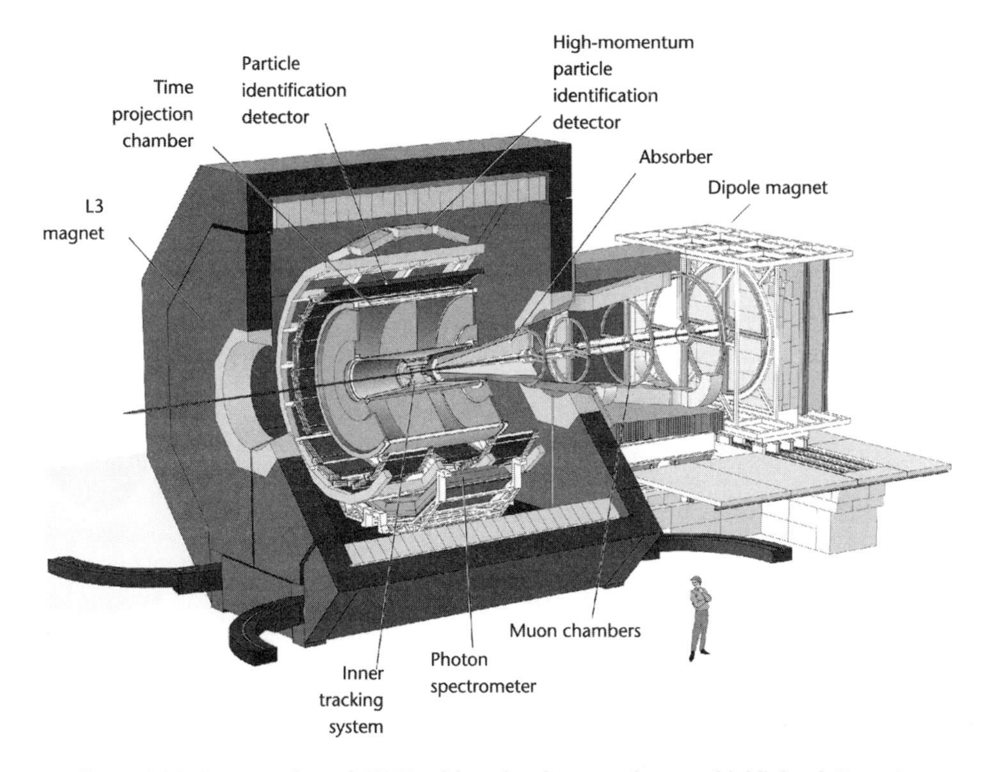

Figure 4.25. Cutaway view of ALICE, with major detector elements highlighted. Note the human figure in the foreground for a size comparison. Figure courtesy of CERN and the ALICE collaboration.

will cross as often as in the proton case (40 million times a second), the lead beams will only collide about 8,000 times per second. When one insists that the lead nuclei hit "head on" and not just a glancing blow, the collision rate will be a paltry 400 times per second, or about a hundred thousand times less often than the LHCb case and 2 million times less often than seen by ATLAS and CMS.

Given these relatively meager collision rate numbers, it might be a good idea to remind ourselves why we want to collide lead beams. It's because when a large number of particles are involved, one can observe different behavior. It's like being at the zoo. As long as each species of animal is segregated in its own enclosure, the animals act in a particular way. But magically erase the cages and mix the animals together, then both types of animals—predators and prey alike—will exhibit behavior not seen before the mixing. When we talk about lead collisions in chapter 7, this will be described in more detail.

In a collision between two lead nuclei, 416 protons and neutrons get smashed into a small volume. With the intense heat of such a collision, it is perhaps unsurprising that many thousands of particles come out of the impact . . . in an extreme case, as many as 20,000. Collisions between lead nuclei are a messy business.

In order to study these complex interactions, the ALICE group designed and built fifteen distinct detectors, using very different technologies. The overarching philosophy of the ALICE detector is not as simple as the cylindrical ATLAS and CMS detectors or the conical LHCb detector. Indeed, ALICE incorporates most of the tricks used in the other three detectors and a few other creative twists as well. Because the level of detail needed to thoroughly discuss the ALICE detector is so great, I've chosen to only sketch the high points of the design.

If you look at figure 4.25, you will see that the detectors near the collision point are a series of cylinders, designed to track particles exiting the collision. Four distinct technologies make up these cylinders. They are a silicon tracker, a detector that utilizes ionization (the time projection chamber, or TPC), a detector that utilizes transition radiation, and a time of flight detector that measures how long it takes for a particle to travel from the collision point to the detector.

Following these detectors are two technologies that don't cover the entire cylinder. These two detectors are Ring Imaging Cerenkov counters and lead tungstate crystals. Surrounding these detectors is a huge solenoid magnet. ALICE has an abbreviated electromagnetic calorimeter (the lead tungstate crystals) and no hadronic calorimeter at all.

On one end of ALICE, one finds a conical style of detector, designed to measure muons. There are other detectors present in ALICE's design, but we'll leave them out in the interest of brevity. Following this short list of technologies, we discuss briefly the various detectors, starting with the first layer, the inner tracking system, or ITS, a silicon-based detector. The ITS sits in a cylinder about 40 inches (1 meter) long and about 34 inches (86 centimeters) in diameter. The system consists of six layers of silicon, with each layer separated by 1 to 3 inches (2.5 to 7.5 centimeters). Not all layers are the full 40 inches (1 meter) long, with the layers closer to the beam being shorter. The inner two layers are silicon pixel detectors, while the outer ones are of a strip type. Altogether, the ALICE silicon detector is comprised of about 12.5 million detectors.

The next layer of ALICE is based on ionization techniques. Charged particles cross the detector, which is mostly filled with gas. The charged particles ionize the gas by knocking electrons off the atoms. Strong electric fields guide the electrons to waiting electronics. By recording which electronics are hit and by measuring when these ionization electrons arrive, the path of the original charged particles can be determined.

This ionization chamber is called a *time projection chamber*, or TPC. It's a large hollow cylinder, with an inner radius of 2.5 feet (78 centimeters) and an outer radius of 9 feet (2.8 meters). The cylinder is 16 feet long (5 meters) with 560,000 electronic channels inside it.

The next detector is called the *transition radiation detector*, or TRD. It fits snugly around the TPC and itself is a hollow cylinder with inner radius of just over 9 feet (2.8 meters) and outer radius of nearly 12 feet (3.8 meters). The cylinder is about 22 feet (6.9 meters) long, and the detector consists of about 1.2 million pieces. The purpose of the TRD is to distinguish between fast-moving electrons and the slower and much, much more common hadrons seen in collisions between lead nuclei.

Next comes a *time of flight detector*, or TOF. As its name suggests, it measures the amount of time it takes a particle to fly from the collision point to the detector. The basic idea is that one uses the particles' velocity to determine their identity. If two particles have the same energy, but one particle is much lighter than the other, the lighter particle will move faster and arrive more quickly. The differences in arrival times are pretty small. Taking two common hadrons (pions and protons with a momentum of 1 GeV/c for those who care), the difference in arrival time is about 3.5 billionths of a second.

The TOF detector is made mostly of gas and works like most ionization detectors. The difference is the electronics, which must include unusually high-tech stopwatches to record the transit time. The TOF detector is very thin and is a shell about a foot (30 centimeters) thick and 24 feet (7.5 meters) long, wrapped around the TRD. The TOF uses about 160,000 elements.

The next two detectors are odd. They don't cover the entire cylinder. They wrap around only a part of the cylinder and are rather short. In both cases, I have the mental image of a blanket on an elephant's back. The blanket neither fully covers the back nor wraps around his stomach.

The reason these detectors don't cover the full cylinder are many. First and foremost, by measuring what one sees in the small area covered by each of these detectors, one can project to what's happening everywhere in the experiment. Another consideration is cost. A final consideration is that the whole detector needs to fit inside the magnet (the next and final layer of ALICE). There wasn't enough room to put the two full cylinders in the remaining space.

The first of these two detectors is the *high momentum particle identification detector*, or HMPID. The HMPID is made using Ring Imaging Cerenkov technology, basically similar to the RICH detectors in LHCb. This detector is intended to determine the identity of highly energetic particles. This complements the TOF, which works best at low energy. The HMPID is composed of 161,000 detectors.

The second detector is called the *photon spectrometer*, or PHOS. Like its name suggests, this detector is designed to measure photons. Consisting of 18,000 lead tungstate blocks, about 1 inch (2.5 centimeters) square and 7 inches (16.5 centimeters) long, this detector has a passing resemblance to the CMS ECAL.

The ALICE detector contains neither a hadron calorimeter nor a muon-detection system in its barrel. In fact, the final layer of ALICE is a large solenoid magnet. This magnet has seen service before. Recall that the tunnel in which the LHC accelerator is built used to house the LEP accelerator. The LEP accelerator hosted four experiments: Aleph, Delphi, L3, and Opal. The L3 magnet is now supplying ALICE's magnetic field.

The L3 magnet has a radius of 16 feet (5 meters) and is 38 feet (12 meters) long. The magnet supplies a relatively modest magnetic field, about ten thousand times the Earth's field and about one-eighth that of CMS. However, this lower strength is exactly what is needed for the kinds of particles ALICE will study.

The last major piece of ALICE is a muon detection system. This is not situated around the barrel, but off to one side, broadly similar to the geometry of LHCb. Basically the muon system consists of five boxes, each box containing what looks like two harps. The harps are surrounded by gas. Muons cross the gas and knock electrons from the gas atoms, and the wires guide the electrons to waiting electronics. These five boxes of wires are combined with a magnet. The magnet bends the path of the charged particles, and the planes of wires measure the amount of bend. From that, the muons' energy is measured. The magnet is about an eighth as strong as CMS, and the muon detection system consists of 1.1 million detectors.

The ALICE detector consists of other subsystems as well, and these detectors play crucial roles in ALICE's discovery mission. However, these detectors have a more technical interest and a detailed description is omitted here. There are at least nine additional small detectors. The ALICE detector can be seen in figure 4.25.

The ALICE upgrade will also be installed predominantly in the second long shutdown period (LS2, 2018 to 2020). Plans are under way to upgrade the inner tracking system and the muon forward (i.e., close to the beam) tracker. This will help reconstruct the interaction vertex. In addition, the electromagnetic calorimetry near the beam will be modified. The HMPID will be replaced by VHMPID (very high momentum particle ID). The time projection chamber will not be changed, but the gas flowing through it will be changed. This will allow for faster readout.

Like all of the detectors, the electronics and triggers will be upgraded. Years

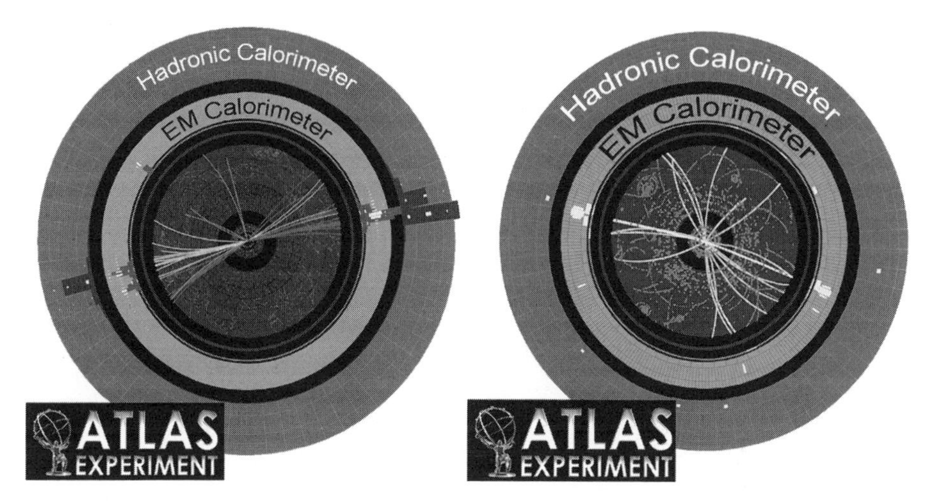

Figure 4.26. These images are event displays—recordings of collisions—from the ATLAS experiment, showing electromagnetic and hadronic calorimeters. The central detectors show the charged particle tracks generated in the collision. *Left,* two quarks or gluons were knocked out of the collision and made jets. *Right,* a Z boson was created and decayed into an electron/positron pair, which are seen in the electromagnetic calorimeter. Note that all of the experiments have similar event displays, but this example explicitly shows the energy in the two separate calorimeters. Figure courtesy of CERN and the ATLAS collaboration.

of operational experience has taught the scientists how to do things better, and this new knowledge is informing the plans for the upgrades.

LHC War Stories

The experiments at the LHC are so large that they take decades to design and build and involve a cast of thousands, including physicists, engineers, computer professional, technicians, and even people to handle the paperwork. With such a large endeavor, it is inevitable that there will be some interesting stories to tell. Most of these stories are told by experienced physicists to younger colleagues after a long day of installing or operating the detector.

These stories are sometimes told over strong European coffee or sometimes a beverage involving ethanol. Depending on the amount of drinks consumed and how long into the night the stories go, they can become embellished to the level of improbability and, depending on the credulity of the audience, occasionally manufactured entirely. One example that I have told more than once is an incident from an experiment I was involved with when I was much younger. In the experiment, we needed to use a lot of steel to shield the equipment from unwanted stray particles. It turned out that an inexpensive source of steel was some old 16 inch (40 centimeter) naval guns from some decommissioned warships. The cylindrical shape of the guns was perfect, and the steel thickness was considerable. The only problem was that the guns had rifling on the inside of the barrels, which left a crack that made for incomplete shielding. As a young graduate student, my job was to crawl inside the guns and fill up the rifling grooves with steel wool to get rid of the cracks. This was hot work in close quarters. It went on for hours, as the packing had to be perfect. More than once I wanted to give up, but with some nudging from my thesis advisor, I grudgingly climbed back into the barrels, packing the steel wool ever so carefully.

After more than 10 hours I was fed up. Tired, sweaty, grimy from the soot inside the guns, I gave up. I refused to go back inside. My advisor told me I couldn't quit, but I assured her that I really did mean it . . . I quit. But then she looked at me very seriously and said, "No, really. You can't quit. Where would I get another student of your caliber?"

I'll let you decide for yourself the veracity of this story.

However, in the pages that follow, I will regale you with some of the tales and facts that have been told about the construction of the LHC detectors. Every one of them is guaranteed to be true. Really.

One such story involves the CMS hadron end calorimeter. It is made of brass. The original design was to use copper, but copper is much more expensive and it turned out that there was a large supply of brass available. This brass came in the form of decommissioned Soviet naval artillery shell casings. These casings were from the World War II era and were no longer considered reliable. The first step required getting permission from the then-commander of the Russian Navy, which turned out to be surprisingly easy. He ordered that battleship shell casings from armories in Murmansk could be used. So the projectiles and propellants were removed, leaving a large supply of high-quality brass. Over a million shells were decommissioned, supplying most of the brass required for the 600 metric ton calorimeters, although the United States did supply copper that was used to make up the remainder. The brass was rolled into plates in Russia and then shipped to Minsk, Belarus, for machining and assembly into the two endcap hadron calorimeters, each weighing 300 metric tons. Figuratively speaking, swords had been forged into instruments of scientific exploration. While it might be an unlikely hope, we can wish that these aren't the last.

The endcap calorimeters aren't the only subject of possible tall tales. The central calorimeters were also made of brass. While the final detector was basically cylindrically shaped, this was accomplished by making eighteen wedges and assembling them together. Each wedge is massive, weighing about 30 metric tons. They needed to be shipped from the assembly plant in Spain to the site of the CMS detector, each one on the back of a truck. One of these wedges had a little adventure. As the truck drove around a traffic circle, what they call a roundabout in Europe, the restraining strap broke, spilling the heavy brass wedge onto the roadway, denting it badly enough that it needed major repairs. You can imagine that pictures of many glum physicists staring at a big hunk of shiny metal on the ground.

Calorimeters might involve a grand tale or two, but the smallest of the detectors are have rubbed elbows with the rich and the powerful. These stories are told around both the CMS and ATLAS "campfires" and involve special dispensation at international borders and first-class plane tickets.

On the scale of the LHC experiments, the pixel detectors at their heart are comparatively small and exceedingly fragile. Tens of millions of tiny silicon detectors are placed with precision nearing 0.00004 inches (a millionth of a meter) and on the thinnest carbon fiber structures that can be made and still be strong enough to support their weight. They're not meant to be shaken, dropped, bumped, or jostled in any way. Yet they were built in locations all over the world, far from their final resting place in the CERN experiments. So how do you get these fragile electronics to Europe?

Well, very carefully is the short answer. The ATLAS pixel detector endcaps were first put in a custom-made holder and then inside a hard-shelled case, lined with thick foam. The case was then attached to the wheels of a baby carriage, a vehicle designed to transport especially fragile and precious cargo. Two physicists wheeled the carriage to the airport and brought it to the security lines where special paperwork had been filed. Two physicists were necessary to make it possible for the couriers to buy food and go to the bathroom and do things like that. Anyone who has traveled solo with a baby will understand the need for two couriers.

The case carrying the detectors would be brought on board the plane as carry-on baggage. However, rather than being x-rayed, the box was opened and inspected and the paperwork carefully scrutinized. Then, the box was brought onto the plane where it had its very own seat purchased. Because of the size of the box wouldn't fit into an ordinary airline seat, the ATLAS pixel detectors flew to Europe first class. One lucky courier sat beside the case sipping complimentary champagne, while the other unlucky companion had to make do with coach. Hey, this research is government funded and money was tight. In order to reassure the other passengers, the couriers brought along photos and information that they handed out in an impromptu session of scientific outreach.

Because the pixel detectors are located closest to the collision point, they experience the harshest radioactive environment and will require eventual replacement. So if you're the kind of traveler who flies first class and you see someone who looks like a scientist sitting next to a carefully strapped down box, you may be seeing another silicon pixel detector heading to CERN and the LHC.

The CMS pixel detectors have a similar story, although without the baby carriage and first-class seats. The forward pixel detectors were put into cushioning foam and surrounded by hard-shelled cases like an attaché case. These cases were hand carried from the Fermilab location in Illinois where they were manufactured to CERN for final assembly. Because the cases were somewhat smaller, they could fit comfortably in a coach seat. So no first-class service for CMS. But, because the boxes didn't have the baby carriage transporter, they were vulnerable to being dropped. The CMS courier joked that if he had dropped them, he'd have just kept on flying. He didn't know where, but he wouldn't return to Illinois.

The plan was to fly to Zurich and drive to CERN, but the complicated paperwork that had been filed only applied to Geneva. As a result, they had to arrange a last-minute connecting flight inside Switzerland. These couriers had to make the trip five times, but it was the least expensive and safest way to transport over 4 million dollars' worth of electronics.

Originally, the builders of the CMS outer silicon detector had similar plans, intending to hand-carry the components on a transatlantic flight from the United States to Geneva, Switzerland, again purchasing the detectors their own seat, this time in business class. However, as we have seen, the events of 9/11 made the paperwork and security checkpoint procedures arduous. Since the outer silicon detector is both bigger and somewhat sturdier, it turned out that it was possible to secure it in well-designed packaging and send it to Switzerland that way.

If the transport of the small detectors came with interesting stories, moving the big stuff had its own drama. Remember that the various pieces of the detectors can weigh hundreds or thousands of tons and be tens of feet (many meters) long. With dimensions of this size, you are probably thinking about big moving equipment, like backhoes or cranes. However, it's important to remember the precision to which these detectors have to be placed and just how full the detector caverns will be. After all, building a bigger cavern means more construction money, and money spent digging is money not available to build the best detectors.

To give you an idea of the stakes, the ATLAS cavern is 150 × 114 × 108 feet (47 × 36 × 45 meters) in size. That's bigger than the size of the detector, but you need to recall that you have to be able to move the various pieces of the detectors away from one another to get access to the equipment. If a device breaks, you have to be able to repair it.

So when the excavation of the cavern was complete, scientists were ecstatic to find that it had been made two inches too big. As soon as the surveyors announced their measurements, physicists began arguing as to which detector deserved that extra bit of space. Keeping in mind that, when the detector is in the operating configuration, the space between the various subdetectors is between 0.2 and 0.8 inches (5 to 20 millimeters), you get a sense of the value of 2 whole inches (5 centimeters).

But before these huge detectors could be assembled at CERN, they needed to be transported from places around the world to Switzerland, which resulted in its own set of fun stories. For instance, there was the time that a driver was so fascinated when he saw the ATLAS liquid argon cryostat crossing the road from the CERN main laboratory to the ATLAS detector hall that he crashed his BMW into the bus in front of him. This is now regarded as ATLAS' first collision. Then there was the pronunciation problem. When an Italian truck driver was told by a French speaker to go to "le CERN," his Italian ear turned that into "Lucerne," a town in central Switzerland. Needless to say, he missed his delivery schedule.

Planning is essential for moving big things, such as the ATLAS endcap magnets. They were assembled in Holland and transported via barge and truck to Geneva, Switzerland. The truck phase began in Strasbourg, France, 240 miles (384 kilometers) from CERN. The path was meticulously scouted in advance to ensure no tunnels, overhead bridges, or any other impediments would be encountered. The trip took five days to complete at a speed of 6 to 10 miles (10 to 16 kilometers) per hour and included the need to dismantle one set of overhead high tension lines at three in the morning.

Each endcap was transported separately. The first trip went relatively smoothly, but when the second one began, the drivers found that a bridge had appeared over the route. It was skiing season in France, and a resort had built the overpass so skiers could just zip from their last run of the day to the hotel for some well-deserved après-ski, rather than slogging across the road. After the French government offered to build an access road for a little over a million dollars, an easier solution was found. A crane lifted the pieces up and over the bridge, while the truck scooted underneath. Problem solved.

CMS's central solenoid magnet is enormous; in fact, it is the most powerful large magnet ever constructed. While money is always a limiting factor on the size of any detector, it turned out that there was a bigger concern. The magnet is essentially a big hollow cylinder, like a log with a big hole drilled along its axis. The maximum diameter of the magnet was 22 feet (7 meters) and was set by being able to transport it through the streets of Cessy, France, a town near the site of the CMS experiment. Transporting the magnet, from Genoa, Italy, where the parts were constructed to CERN, where they were assembled, was a spectacle. That trip took 10 days. Then, later, when the huge cylinder was ready for installation, the magnet then travelled the final 5 miles (8 kilometers) to the detector hall. No houses were scraped in the process, although there were a couple of times when the onlookers held their breath.

Stories like this are part and parcel of the lore of any large project. They are passed down from generation to generation, from thesis advisor to student and then again as the student becomes the master. I have even heard stories told of Enrico Fermi, although he passed away before I was born. No good story ever dies; it just gets recycled, repurposed, and told again.

Why did I ever tell you when I . . .

What's Next?

We have now learned something about the four major detectors at the LHC. These four detectors began taking data in 2010 and have already run through the first months of 2013. They will be at the forefront of physics research for at least the next two decades. It's impossible to predict which of the four detectors will make the crucial observation that reveals something entirely new.

Stay tuned!

5

TEETHING PAINS AND TRIUMPHS

Outside the aluminum walls, the frigid air whistles by. The whining engines drone their monotonous tune, until their oppressive voice is only felt and no longer heard. Lacy white clouds far below obscure the sullen northern Atlantic waters, while the drink cart clinks by, dispensing much-needed caffeine as groggy people begin to rouse and decide what time their body thinks it is. Overhead, the string of monitors signal that we will soon be above Ireland and that it won't be long before again we'll be descending to terra firma. Somewhere ahead, over the gently curving horizon, the Geneva airport beckons.

When I give my public lectures, one thing the audience often comments on is the glamour of international travel. They imagine exotic places: wine and Rome, cafes and the allure of Paris. And I admit that I was once enthralled with the travel that is part of the life of a particle physics researcher.

But, as any frequent business traveler will tell you, the idea of globetrotting quickly loses its cachet. After the twentieth trip in as many months, the prospect of another 10 hours on a plane, the missed anniversaries, soccer games, and dance recitals, it all begins to pall. Gold membership with half a dozen airlines can ease the pain, but only so much.

With the drudgery of frequent international travel, you might ask yourself why we scientists do it? Why subject ourselves to the frequent absences from home? What could possibly make it worthwhile?

There can be one but answer to this question: the excitement of discovery. As one clears customs and heads to the taxi stand, the mind leaps forward to the laboratory. CERN hosts the highest energy particle accelerator in the world. It is there where the frontier of knowledge is being explored; there that the universe is grudgingly giving up its secrets. For a particle physicist interested in the energy frontier, it is the place to be.

For those of you who have not yet visited CERN and its environs, let me be the first to invite you. Geneva is located on the western end of Lake Leman, where

Figure 5.1. The main CERN site is located astride the French and Swiss border; however the LHC accelerator extends much farther than shown here. Figure courtesy CERN.

the Rhone River exits and winds its way across the French countryside. Geneva is a pretty place, nestled between the Alps to the south and the Jura Mountains to the north. If you come during the winter months, there are tremendous skiing opportunities nearby, while the same mountains provide delightful hiking during warmer times. A romantic walk along the lake will pass the breathtaking Jet d'Eau, shooting the lake's water more than 400 feet (140 meters) in the air. And then, of course, there is the chocolate.

But if you come here, be sure to visit. CERN welcomes visitors and, if you make an appointment, they can arrange a tour for you. Luckily for science enthusiasts, if the accelerator is not running, it is possible to visit the LHC experiments and get a sense of the feel of the place. Figure 5.1 is an aerial view of the facility. Industrial buildings sprawl over the campus, each with a purpose to advance CERN's scientific mission. And, as you readers now know, what a mission it is!

As this book goes to press, the LHC is temporarily shut down in order to make repairs and upgrade both the accelerator and the detectors. The period of 2010 to 2012 saw the LHC operating spectacularly well, although at a collision energy of 7 or 8 trillion electron volts, or TeV, considerably below the design 14 TeV. This lower energy was clearly not intended from the outset, and there is a story in how that came to be.

Extraordinary Excitement

So let us look back in time to September 2008. Sarah Palin had just been selected as a U.S. vice presidential candidate, beginning a national presence that has waned but not yet ended. Hurricane Gustav had just passed over New Orleans without leaving the mark of the earlier Katrina. And particle physicists were eagerly anticipating the turn on of the LHC.

At that time, Lyndon Evans was the project manager for the Large Hadron Collider. This Welsh scientist had spent the previous 14 years involved in the design and construction of the LHC and was now wrapping up his term as the leader of this 8-billion-dollar technological marvel. He was concerned with the impending turn on of the accelerator. Evans wasn't worried about black holes or other silly doomsday scenarios bandied about by irresponsible media outlets. He and the other CERN scientists knew better. He was worried whether the LHC would work at all.

Well, that's not quite accurate. Evans led a masterful team of scientists, engineers, and technicians. There was no question that the LHC would work, at least eventually. He had commissioned other accelerators before. He knew the drill. However, this stage of the commissioning of the LHC was going to be different. It was going to be done live, on international TV, in front of an audience of literally billions of eyes. With no dress rehearsal, the LHC accelerator operations team was going to thread a beam of protons a fraction of an inch (a few millimeters) in diameter, through a 17 mile (27 kilometer) long pipe that was just an inch or two (a few centimeters) in diameter. No pressure at all.

The CERN management was confident that Evans's team had done their job properly. They had alerted the world's press, and the media had responded. While hundreds of media outlets looked on, on the morning of September 10, 2008, the accelerator crew started their work.

While CERN hosted hundreds of reporters from all over the world, the action was not only at CERN. At universities and national laboratories across the globe, live action broadcasts were available. My own laboratory of Fermilab is in Illinois, which is 7 hours behind Switzerland. The first attempt to steer the beam around the LHC began at 2 a.m., Chicago time. In order to entice people to attend, we had a themed party where everyone was invited to wear their pajamas. Pierre Odonne, then the Fermilab laboratory director, oversaw the festivities while clad in oversized footed PJs. One respected scientist attended in an outfit that would have not looked out of place on Hugh Hefner in his heyday, complete with smoking jacket and cravat. More than 400 people, including a hundred in various types of nightwear, attended the festivities. Local school kids showed up

Figure 5.2. The LHC control room was packed on the day of first beam. In this photo, CERN director Robert Aymar (*center left*) and LHC project manager Lyn Evans (*center*) orchestrate the effort. Figure courtesy CERN.

to see the historic event, joined by representatives from the U.S. funding agencies and, of course, the national press.

Meanwhile, back at CERN, the media and the scientists waited. The first goal was to thread a single beam around the ring. Prior to this attempt, the LHC accelerator operators had only steered the beam around a fraction of the ring. It was show time.

A little after 9 a.m., CERN time, the beam had been brought to the entrance of the LHC. There are eight places around the ring with no magnets and with straight beam pipes. These are places where, in principle, detectors could be housed. One by one, the beam was guided to each detector and the magnet settings checked. Earlier, there had been a nervous moment when the cooling for some magnets looked a little unstable, but a little attention from the technicians settled it down. In under an hour, the beam was brought full circle, around the 17 mile (27-kilometer) long ring. Applause rang out across the world. Robert Aymar, then the director general of CERN, made a statement, and congratulations from the other laboratory directors poured in. Evans' team had done it. Figure 5.2 gives a sense of what it was like to be in the control room that day.

However, the LHC team was not done. They had thus far only shaken down a single beam. The LHC consists of two beams, in separate beam pipes. While the world partied, the operations team started circulating beam going the other direction. This effort was also successful.

The job was clearly not complete. In fact, what the LHC team had managed to do was to circulate beams with tiny numbers of protons and only at a low initial energy. The beams were metaphorically as broad as a barn and hadn't been squeezed down to their final pinprick size. While the LHC was designed to operate with beams of a full 7 TeV, those entering the accelerator were at the much more modest energy of 0.45 TeV. The LHC's purpose was to accelerate the protons to the final energy. Further, in order for the "C" in the LHC's name to be accurate, the beams would need to collide. That required two counter-rotating beams, going in opposite directions and preferably at maximum energy and at an extremely narrow width. All these technical milestones had yet to be achieved.

But September 10 was not intended to be the last word. The beam commissioning stage of such a large instrument necessarily can take weeks. The press was invited to see an important step but not the final one. Still, the public awareness was extraordinary. Briefly, high-energy physics got the kind of attention one normally associates with one of NASA's grand achievements. The landing on the moon got more attention, but for particle physics, this was unprecedented. Particle detectors graced the cover of magazines like *Newsweek*; stories on CERN got above-the-fold coverage in leading newspapers and were the top story for many a national news broadcast. Even that most modern of cultural affirmations, a special Google doodle appeared (figure 5.3).

As far as the world was concerned, the LHC had arrived. After all, a press conference had been held; champagne had been drunk on live TV. Physicists had been clear in their statements of what had been achieved but, in the excitement, it is easy to see how the misunderstanding came to be. As the television crews were loading their trucks, the accelerator physicists returned to work.

First the beams needed to be "captured," which simply means that the accelerating radio frequency (RF) voltage (discussed in chapter 3) needed to be timed precisely so that the beam arrived at exactly the right moment to be accelerated. Shifts in time of a few billionths of a second would destroy the machine's ability to bring the protons to operating energy. Timing in the beam can often take days, but the operations team took only hours. Indeed, as the last of the satellite trucks pulled out of the parking lot, this important milestone had passed. Figure 5.4 shows how physicists knew they had succeeded. In the first panel, the beam had just been turned on. In the second panel, we were able to see that the beam's arrival had been properly synchronized, meaning the amount of beam was going to be consistent.

Figure 5.3. This logo appeared in Google for all countries on the day of first circulating beam, just one example of the universal appeal of the LHC and its search for the Higgs boson. Google logo © Google Inc. Used with permission.

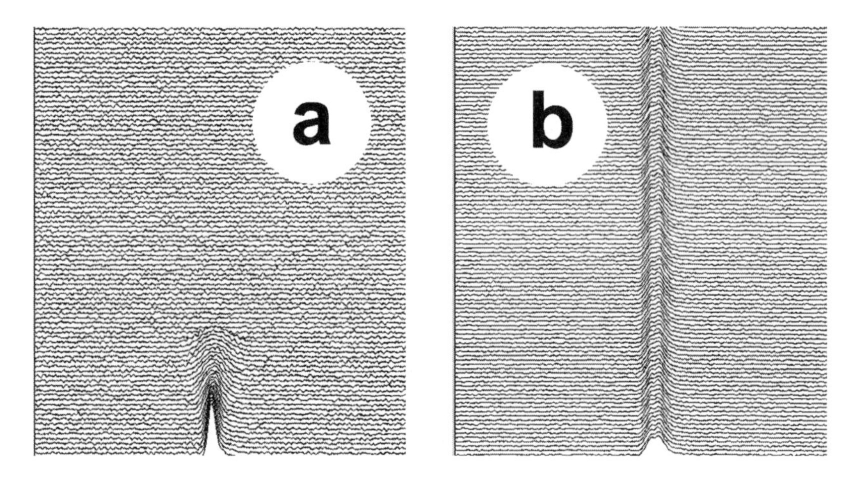

Figure 5.4. The so-called mountain range plots show how well the beam was timed. The horizontal axis is the time within a beam orbit. Each horizontal line in a plot is a single orbit. The height of the bump shows how much beam there is. *a*, First beam, with the accelerating radio frequency (RF) turned off. After a few orbits, the beam is lost. *b*, RF on and timed in properly. We see that the beam is in the same spot, orbit after orbit, and the amount of beam stays constant. This is the correct configuration, and it was achieved at the LHC in a scant few hours of operation. Figure courtesy of CERN.

The next big step was to bring the magnets up to operating current. As the energy of the accelerator increases from the beam injection energy to the maximum design, the magnetic field in the magnets needs to be increased. This is done by increasing the electrical current in the magnets arrayed around the ring. In this step there was considerable danger.

The Incident

The LHC has eight named points around the ring at which collisions can occur, at least in principle. They are numbered one through eight. Between these points are eight long sectors of magnets, labeled by the points that surround them, sector 12, 23, and so on. Figure 5.5 shows a schematic of the accelerator. Prior to the successes of September 10, all of the sectors, save one, had been tested with electrical currents that would contain a beam of 5.5 TeV. Sector 34 had only been tested to 4.2 TeV. All sectors needed to be tested at currents for 7 TeV running. And then a serendipitous problem occurred. On September 12, just shy of midnight, an electrical transformer broke near point 8. This transformer was part of the electrical system for cooling the magnets in sectors 78 and 81. This is no small transformer, weighing in at 30 tons, and the electrical crews spent Saturday and Sunday replacing it. The temperature of the strings of magnets had risen significantly, and some time was required to cool them back down to their operating temperature of –456°F (–271°C). While the cool down was ongoing, the operational crew decided to do some more tests of the unaffected sectors. They decided to ramp up the magnets in sector 34 to 9,300 amperes of current. This was well below the final operating point of 11,850 amperes and the eventually intended "over current" test of 12,400 amperes.

Things were apparently progressing beautifully, with the electrical current slowly rising. And then the current reached 8,700 amperes. This is when things started to go bad. Quickly. The alarm board went red. The fire alarms in the LHC tunnel 300 feet (100 meters) belowground went off. The power to the entire sector was automatically cut off. Aboveground, in the control room, pandemonium, shock, and initial confusion ensued. It looked a lot like the scene in the movie *Apollo 13* just after the explosion in the oxygen tank. Only a few things could have made the control system go haywire like that, and none of them meant anything good.

I was in Illinois on the day of the incident. I heard about it as soon as the accelerator operators had something to report to the experimenters. The U.S. CMS group has a weekly meeting at 11 a.m. on Fridays. We sat stunned during that meeting as the news was announced. Indeed we learned about the problem so soon after it occurred that there had not been time to take stock of just how bad the damage was. It was clear it was bad. But how bad?

Over the ensuing days and weeks, the situation was clarified. CERN commissioned a postmortem study of the incident that included experts in all manners of accelerator construction in fields such as cryogenics, mechanical engineering, and electrical engineering. It was critical to get to the bottom of things and

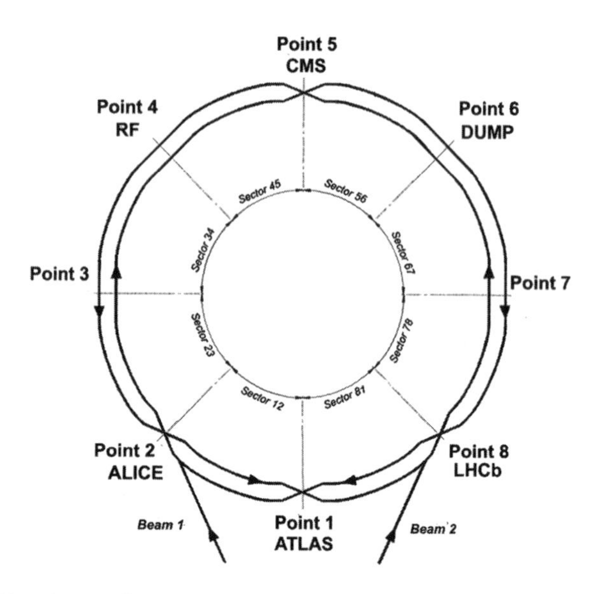

Figure 5.5. The LHC consists of eight major sectors, separated by points at which collisions can occur. Sector 34 is the one that was damaged in the fall of 2008. Figure courtesy of CERN.

to find out what had occurred. Was the problem a broad design flaw, an incidental (although devastating) problem, or what? The task force issued many progress reports, but it was not until March of 2009 that the final report was released.

We can now reconstruct what happened on that fateful day. Each and every magnet had been tested aboveground to full operating conditions. Further, each magnet was heavily instrumented against electrical and cryogenic faults. If there was a problem with the electrical or cooling system, circuitry would detect the problem and react in fractions of a second to shunt the electricity out of the magnets to protect the system.

However, there was one weakness. These magnets needed to be electrically connected into a single large circuit. This was accomplished by taking the superconducting cables from one magnet and soldering them to the next. These superconducting cables were buried in large copper bus bars as a safety measure. In addition to providing physical stability, if the superconducting cable ever got warm enough to stop being superconducting, the heavy-duty copper would take over. But, in a technical oversight, the safety circuitry did not extend to these soldered splices. Naturally, this was where the problems all started.

Understanding what had gone wrong was difficult. The damage to the magnets was severe, with large sections of conductor having been simply vaporized. However, the forensic engineers have been able to reconstruct what occurred.

Essentially, the soldering connection between two magnets was extremely poor or even possibly nonexistent. As the electrical current increased, a resistance developed between the two magnets, causing overheating. In fractions of a second, the temperature became hot enough to first melt the solder and then melt the copper bus bar. The electrical current had nowhere to go, and an electrical arc developed. An electrical arc is like the little spark you might see when you plug in an appliance; except this spark had the force of nearly 10,000 amperes—remember this is the equivalent of a hundred households—of current behind it. If this had been the worst of it, the situation would have been recoverable. However, things were about to get worse.

Remember that the LHC magnets are superconducting, which means they are cooled by liquid helium. This liquid helium is stored inside what is essentially a huge Thermos bottle inside the magnets, and the electrical arc burned a hole into the bottle. That spelled real trouble. When a liquid turns into a gas, it expands to about 700 times its volume as a liquid. Over the course of a couple of minutes, two tons of liquid helium were released. When it expanded, it pushed the regular air down the tunnel. This totally displaced the oxygen in the tunnel for a distance of about a mile (1600 meters) in both directions. The initial speed of helium "wind" was about an Olympic-class runner's pace of a four minute mile (400 meters per minute). You couldn't outrun it and when it passed you, there was no air left to breathe. The air didn't return to normal for more than 5 hours.

This outpouring of helium wasn't gentle. The pressure of the expanding helium jetted out of the magnets. Even though some of them weighed as much as 35 tons, they moved a foot or so (many tens of centimeters). This movement ripped the supports out of the floor, shattering the concrete in the process. It also either squashed or tore apart the electrical leads, coolant lines, and other connections on the opposite side of the magnets. This damage set up additional electrical arcs and the release of more helium. Figure 5.6 shows some of the damage. The helium got into the beam pipes and blew soot and insulation over huge distances. Eventually 16 tons of liquid helium were lost. Finally, remember that the helium had been chilled to −456°F (−271°C). Movies taken hours after the incident showed thick coverings of frost on the magnets for huge distances from the original electrical arc. This was a spectacular incident. Luckily, CERN's strict safety regulations mandate that nobody be near the accelerator when it is operating, and no person was ever in any danger.

As our understanding of the root cause of the incident improved, it became ever-clearer that we needed to proceed carefully. Detailed measurements of the splices revealed that to run safely at full energy would require that all of the magnet splices around the LHC ring would have to be redone. This is a big job

Figure 5.6. The damage to the LHC was considerable. *Top*, a vaporized copper conductor. This damage is localized, although there were several instances like this. *Bottom*, damage to the connections between two magnets weighing tens of tons with the flexible cover off (*left*) and on (*right*). The helium released moved the magnets several tens of centimeters and distorted all of the delicate interconnections. Figure courtesy of CERN.

and would necessarily take a year or more to accomplish. So, naturally the question then became whether the magnet splices were sufficiently strong to be able to work at lower electrical currents. That question is the subject of its own dramatic tale.

The Reaction

The first official conversations began at the Chamonix workshop in early 2009. This workshop, held in the shadow of the imposing Mont Blanc, is an annual getaway for the CERN leadership to assess the state of affairs at that organization. In an era dominated by the LHC, it is natural that the discussions were focused on that topic; especially after the unfortunate incident of September

2008. Representatives from the experiments, as well as from the various national funding agencies, were present.

Philippe Lebrun was the chair of the task force investigating the incident and presented the summary report. However, there were many other, more detailed reports by the various experts. There was a lot to absorb, especially for the new CERN director general, Rolf-Dieter Heuer. He had taken over just a month previously from Robert Aymar. While the earlier directors presided over the construction of the LHC, it was Heuer who would have to navigate the future waters. Many options were discussed, including whether it made sense to run with lower energy beams. No decision was made at Chamonix, but a clear path was laid to better investigate the options. It would take more information for Heuer to make a decision.

Meanwhile, back at CERN, the technical teams were continuing to effect repairs on the damaged equipment. The damage extended over 2,400 feet (750 meters). Ultimately fifty-three magnets, each weighing tens of tons, were removed for cleaning and repair. Eight others were cleaned in situ. But even the undamaged magnets, far from sector 34, required attention. Cautiousness dictated that all of the solder splices in the accelerator be tested. Nobody wanted to chance another incident. Further, many improvements were undertaken to mitigate the effects of any possible future problems. Pressure relief valves were improved on many magnets, and instrumentation was added to sense and respond to trouble.

Stories of the LHC's woes were common in the press and even occasionally became comical. A mundane incident in a power station that caused the electricity to go out was reported as being caused by a bird dropping a piece of bread into the transformer. The evidence for the "saboteur bird" wasn't particularly strong, but it made for an excellent story and so made it into the blogosphere. Then there was the theoretical physics paper that said that perhaps the universe didn't want the Higgs boson to be discovered and, when it was discovered some time in the future, this caused a rift in the space-time continuum. This rift somehow involved some kind of time travel to keep breaking the LHC to make sure the Higgs boson wasn't discovered after all. The paper was more sophisticated, but that's how it was reported by the media. The press had a lot of fun with both stories, and it was common to have to answer questions about these sorts of things in my lectures. I didn't mind so much since it at least indicated that the public's interest level was high.

The LHC repairs were finally completed, and more than a year after the incident, on November 23, 2009, circulating beams returned. These beams were at the low injection energy of 0.45 TeV and with very small numbers of protons. The first handful of collisions occurred in the ATLAS detector in the afternoon.

The other experiments followed suit later that day. A week later, on the evening of November 29, a single beam was accelerated to 1.05 TeV, beyond the record held by the Fermilab Tevatron. Three hours later, just after midnight, both beams were accelerated to 1.18 TeV. With that achievement, the LHC quietly became the highest energy accelerator in the world.

However, as I always remind my accelerator colleagues, accelerating particles without recording the collisions is somewhat pointless. The experiments needed to record data.

As mentioned above, a few collisions occurred on these last days of November. But it was on December 6 that the accelerator intentionally collided beams of protons at 0.45 TeV. Each experiment eventually recorded about a million collisions. On December 12, the beams were ramped to 1.18 TeV, and, again, collisions were recorded. The beam intensity and energy were quite low, but these data were invaluable as a means of gaining operational experience. In addition, the experiments were able to use these data to publish some measurements in professional journals. ALICE published first, followed by CMS, and then ATLAS. Perhaps the most noteworthy aspect of the publishing process was the incredibly short time it took. In just a few short weeks, the papers were submitted. Usually it can take many months to analyze data and get it into good enough shape to release. Even though the physics message of the papers was modest, this boded extremely well for future efforts.

The Drought before the Deluge

The LHC shut down for the Christmas holidays on December 16, 2009. While the outside world took little note, the CERN community celebrated, as the brief running period was a resounding success. Christmas had come early, and Santa was especially generous that year.

The accelerator technicians and operation managers of the various experiments took the months of January and February to fix small problems that had presented themselves during the running in December. These repairs were minor and not a source of concern for scientists, for they knew what was coming. It was time to seriously start collecting data.

The annual CERN program meeting in Chamonix in early 2010 had finally resulted in a decision on the beam energy. The studies of the various splices suggested that it was probably necessary to replace them all if the goal was to run at the full design energy of 7 TeV per beam. However, it would be safe to run at half energy; 3.5 TeV per beam, maybe even a bit higher. At Chamonix, it was decided to take the conservative approach and not go above 3.5 TeV until proper repairs were completed. If enough data could be collected at this energy, there were still good prospects for new physics discoveries. It also would give time for the experimental collaborations to shake down their equipment, software, and indeed even respective social and management organizations. Experimental particle physics collaborations consisting of 3,000 physicists had not been seen before. It was inevitable that there would be some kinks in the process. The decision to run the LHC for 18 to 24 months at a beam energy of 3.5 TeV per beam would serve many purposes. At the time, the plan was to turn the accelerator off in late 2011 to undertake the repairs necessary to operate at the full design energy of 14 TeV. We will see that this plan changed.

So with the operating energy decided, the LHC was powered again on February 27, 2010. It came up smoothly, with beams of 0.45 TeV established rather quickly. The real question was whether beams of 3.5 TeV could be reliably achieved. There were some last-minute anxieties about the process and the decision was taken to accelerate over the course of 75 minutes rather than the design time of 15 minutes. This was from an abundance of caution. But, just before dawn on March 19, two counter-rotating beams were accelerated to 3.5 TeV each.

I happened to be at CERN that particular day and, of course, was awake for the accomplishment. Over the course of an hour and a quarter, I sipped espresso and watched the beam energy creep upward. My colleagues and I chatted back and forth, wondering whether the beams would make it. A measure of our confidence in the accelerator crew is that nobody was willing to bet against them. When it reached the intended energy, the room erupted in cheers, but I had to sit back and marvel. We had done it. A new energy frontier was about to be explored. But not quite. The beams had intentionally not been collided. That would take a few more days.

Naturally the CERN management wanted to share their success with the world; possibly to redeem themselves in the world's eyes. They notified the international media that the time had come to turn the accelerator over to the physicists who were wriggling like a puppy at the prospect of new data. To mark the occasion, the press was invited to witness the first collisions between protons at 7 TeV. Like September 10, 2008, the world's media converged on Geneva, Switzerland. Nobody wanted to miss the excitement. And, as the world looked on, on March 30, just after lunch time (and after two nerve-wracking false starts), it happened. The beams were crossed at operating energy and collisions were observed. Two such collisions are shown in figure 5.7. After decades of work, involving tens of thousands of people, the LHC was working. Data were flowing, and discoveries could be just around the corner.

With the successful collision of two beams of protons with an energy of 3.5 TeV each, the prologue of the LHC was complete. Figure 5.7 shows some of the first collisions that brought relief and elation after the collider had been repaired. No matter that it had taken decades to get to that point, this was merely the end

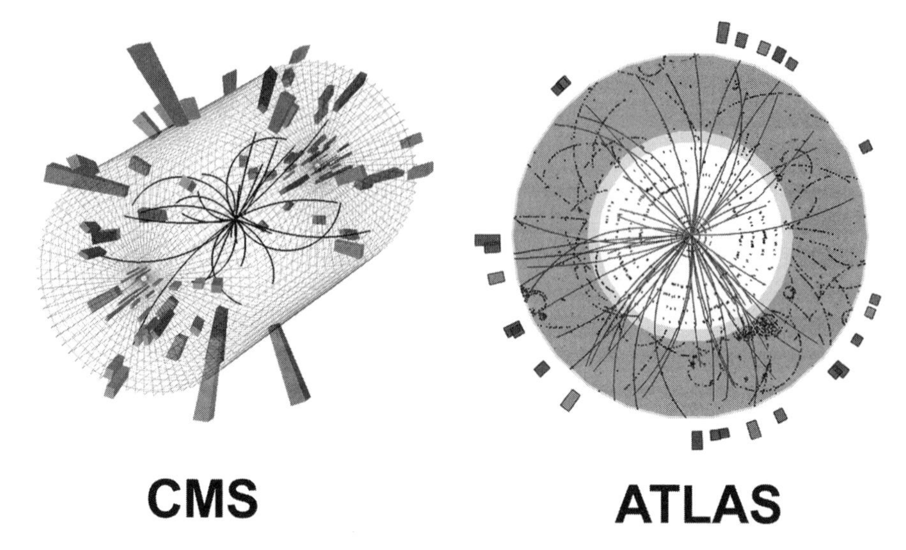

Figure 5.7. On March 30, 2010, the LHC accelerator began delivering collisions at the unprecedented energy of 3.5 trillion electron volts per beam. These event displays are from the very first collisions recorded by two of the LHC experiments. By now there have been billions of collisions recorded, many far more interesting than these, but these are the first. Figure courtesy of CERN.

of the beginning. In the same way that the final nails pounded into Columbus' three ships signaled the completion of the first stage of his history-making voyage, it was the journey ahead that excited. Like the explorers of yesteryear, today's physicists were poised to leap off into the unknown, eager to see things never seen before, and to perhaps discover things so bizarre that they were never before even imagined. The quantum frontier beckoned.

The Fun Continues

The year 2010 was a great one for the LHC, although the beam was still really quite modest. However, the 7 TeV of collision energy did allow the LHC scientists to explore an entirely new energy realm, much higher than had been achieved before. While it is natural that a few intrepid explorers would try to get a jump on their colleagues and see if a quick discovery might be had, the fact was that only a few discovery-oriented analysis topics could be pursued with so little data. Certainly a light-mass Higgs boson wasn't in the cards, although a heavy one was within the LHC's discovery reach. (In an oddity of physics, a heavier Higgs boson is easier to find than a lighter one. This is because the way in which a Higgs boson can decay depends on its mass. A light Higgs boson decays in a nondescript way, while a heavy Higgs boson has more easily identifiable features.) But mostly 2010 was a shakedown year. Scientists rediscovered the standard model at this higher energy. Quarks and gluons were scattered, jets and electrons and muons were observed. Researchers were able to see W and Z bosons, top quarks, bottom quarks—the whole panoply of subatomic particles that had long inhabited the textbooks.

Even though this might sound like a pedestrian endeavor, it really wasn't. First, there was always the chance that between the 2 TeV collision energy of the Tevatron and the 7 TeV energy at which the LHC was running that the rules could have changed tremendously, like if a person is reaching a higher and higher altitude by climbing a mountain. For a long time, taking one step after another gets the hiker to a higher elevation. However, eventually the climber will reach the top of the mountain and what worked before (taking one step after another) will no longer allow the alpinist to increase his altitude. If the rules changed at (say) 5 TeV, it could well have been that exploring such a tiny bit of data at 7 TeV could have blown everybody's mind.

As it happened, the rules didn't change. This wasn't a surprise; nobody expected that they would. The conventional standard model physics were still important. If the LHC experiments couldn't identify and study things long known about like top quarks and W and Z bosons, why would anyone believe researchers who started claiming discovery of things never before seen? Thus 2010 was a very valuable year, even if it didn't result in any huge discoveries. There were some little surprises that kept the enthusiasm level high, one of which came late in the year when the LHC spent a month colliding lead nuclei together. (We'll talk more about the physics of scattering heavy nuclei in chapter 7, but collisions of lead nuclei are very messy, with 416 protons and neutrons involved.) The temperatures of these collisions are very hot—so hot that it is thought that the protons and neutrons could "melt" and quarks and gluons could run around willy-nilly. If that's true, then we'd expect to see a change in the nature of collisions.

To get an idea of what I mean by this, let's talk baseball. A third baseman can throw a ball to the home plate with a pretty good chance of the ball getting there. But, if the field was covered with a huge crowd, it is harder for the ball to make it to the catcher. The ball is likely to bounce into someone and not hit its intended target. Similarly, if you smash a quark or gluon out of a collision, it should easily escape. However, in a collision between lead nuclei, the scattered particle has to plow through the crowd of melted nucleons. You'd expect to see fewer quarks and gluons scattered in these collisions between heavy nuclei. Earlier measurements at the Relativistic Heavy Ion Collider on Long Island in New York State had indirectly supported the expectation that the subatomic crowd would reduce the numbers of quarks and gluons coming out of the collisions, but ATLAS was able to use the 2010 data to make the first direct observation of this phenomenon.

If the year 2010 was one of no discoveries but valuable validation of the accelerator, detectors, algorithms, and even sociology of such huge experiments, 2011 was a year for which the hopes were much higher. The intensity of the beams delivered to the experiments was greatly increased. We'll talk about the exciting search for the Higgs boson in the next chapter, so we'll skip that discussion here, except to say that 2011 was the year in which the LHC unambiguously took over from the Tevatron. Given the fact that the LHC had 3.5 times the energy of the Tevatron and even brighter beams, the transition was inevitable. The decision to shut down the Tevatron at the end of September 2011 was a sad period for those who had spent years and decades exploring the unknown using its data. But regrets like that are temporary, and many of us moved on to the shiny new toy that was the LHC.

In 2011 a little more than a hundred times as much beam was delivered as in 2010. The extra beam allowed for much more precise measurements of standard model physics and a serious push to find the Higgs boson if it existed. The year 2011 foreshadowed the excitement of 2012 that is described in the next chapter.

At the 2010 Chamonix meeting, 2012 was envisioned as the year during which the LHC would be repaired, allowing operation at the full 14 TeV collision energy. But plans changed. The LHC's rival of sorts, Tevatron, was starting to see an "interesting" excess of events in their Higgs searches that could well mean that they were beginning to see the Higgs boson; nothing worth panicking over, but enough for the LHC to take notice. The LHC might have been shut down for a year and a half to 2 years. If the Tevatron beat the LHC to the punch in those long months, a mob carrying torches and pitchforks might have descended on the CERN director's office. Thus it was decided to extend the LHC running for a year. The accelerator was running superbly, and allowing it to run for another year would give the experimenters enough data to find the Higgs boson if it existed. In fact, the CERN director even committed to extending the 2012 run for a couple of months if it turned out to be necessary to go the final distance.

The LHC turned on for real on April 5, 2012, and it was immediately clear that 2012 was going to be a bumper year for data. Everything about it beat the conditions of 2011 by a significant factor. To give a sense of just how much better things were, both ATLAS and CMS recorded as much data by July 2012 as in all of 2011. By December of 2012, that year's haul was about four times as much as 2011.

In addition, the experience of running the accelerator in 2010 and 2011 had given the operators considerable confidence in the stability of the LHC. They advised the director that 7 TeV was actually a conservative operating point. The LHC could run at 8 TeV with no credible danger. While raising the energy meant that the CMS and ATLAS researchers would have to revisit vast amounts of their analysis techniques and simulations, the extra energy would give at least a 20% boost in the number of Higgs bosons produced. Additional work is never much fun, but the enhancement in discovery potential made the decision a no brainer. Both groups voted for the energy increase, and the CERN director agreed.

There was one tricky and subtle consequence about running the LHC at higher energies and brighter beams. You'll recall from chapter 3 that the LHC beams are actually a series of bunches. These bunches can be envisioned as looking like little sticks of spaghetti, about a foot (30 centimeters) long and actually much, much thinner, perhaps more like a long, thin, straight hair. Each of these bunches contain perhaps 100 billion protons, and there are hundreds or thousands of these bunches in the accelerator, circulating dutifully like so many train cars. While the original design of the LHC had the bunches separated by 25 nanoseconds of flight time, this turned out to be harder to do that was originally anticipated. Because of this, the beam was run in "50 nanosecond mode." This means that the bunches were separated by longer distances. If you think about this for a minute, you'll realize that if the beam was originally intended to be dispersed in bunches separated by 25 and they were instead separated by 50 nanoseconds, then there were half as many bunches in the accelerator as designed and therefore each bunch had twice as many protons.

The consequences of that are that as two bunches cross each other inside the detectors, there are twice as many collisions than expected and therefore twice as much of a mess. Any physics collision revealing new physical phenomena could be accompanied by twenty or fifty or even more simultaneous collisions. Learning how to deal with this environment occupied a lot of the researchers' time in 2012.

The 2012 run ended with 2 months into 2013 colliding a proton beam with a beam of lead nuclei. Combining these new data with the huge set of data studying proton-proton collisions and earlier collisions between pairs of lead nuclei provided additional insights into how the strong force we learned about in chapter 2 works in the crowded environment of hundreds of protons and neutrons plowing into one another.

Future

So where do we stand? The LHC stopped colliding protons in December of 2012 and collided protons and lead together for a couple of months early in 2013. By mid-February of 2013, the accelerator shut down to effect repairs on the connections between magnets. In addition, small improvements were undertaken by the four experiments.

As of this writing, the plan is that in the fall of 2014 the repairs will be completed. The LHC will be cautiously turned on beginning around then. After a short period of shakedown, CERN will again take its extended Christmas holiday and about early March 2015 (give or take), the LHC will turn on again, this time at least 12.5 TeV of energy and probably 13 and perhaps even as high as 14 TeV. (If, when you read this, the energies and dates turned out to be a little wrong, let me remind you of Yogi Berra's cautionary quip that it is hard to make predictions, especially about the future. However, these predictions are expected to be reasonably accurate.)

Over the next couple of years, the amount of delivered beam will be perhaps three times that seen over the entire 2010 to 2012 running period. More data and higher energy means new frontiers to explore. Nobody can predict what will be found, but it doesn't take a psychic to know that it is going to be exciting.

6

The Dramatic Higgs Saga

In the United States, the Fourth of July is a time for celebration and fireworks. In 2012, the fireworks were of a scientific, rather than a pyrotechnic, variety. As many Americans slumbered, a scientific seminar was taking place in western Switzerland, just a few hundred yards (meters) from the French border. Scientific presentations taking place in the CERN main auditorium are not news; I have attended more than I can remember. Over the course of its existence, the main auditorium has hosted thousands of seminars and about a half a dozen presentations that had scientifically historic implications. The discovery of 2012 is easily in the top three and arguably number one. After a hunt spanning nearly half a century, it looked like the last undiscovered component of the standard model might have been found. So what was uncovered and why did it take the Large Hadron Collider to find it?

Forces in Search of a Field

In chapter 2, we talked about the standard model, which describes the behavior of all the familiar matter in the universe, the kind that makes up you and me—everything from atoms to galaxies. Take the quarks and lepton building blocks, and we can construct every point of light you see in the night sky. With the inclusion of three forces, the standard model can even explain how the stars burn and how we can see them over incomprehensible distances.

When people talk about forces, they usually mean something that changes an object's motion, like pushing a car out of a ditch or popping open the parachute of a drag racer. This meaning still applies in particle physics; as when we collide two particles together, they can bounce off one another and head in different directions. However, in particle physics, a force has a more general meaning and might better be called an "interaction." In addition to the phenomena we intuitively understand to be a force, in the particle realm forces can also cause particles to decay and transform into other particles.

In chapter 2, we discussed the nature of the weak, strong, and electromagnetic forces. (We also discussed a fourth force—gravity—but it is not relevant here.) But it's a reasonable question to ask whether these three apparent forces are actually distinct or if they are actually different "faces" of a single, unified, phenomenon. After all, history is full of cases where seemingly disparate phenomena weren't different at all. For instance, it wasn't until Isaac Newton came up with his theory of universal gravitation that we realized that the fact that things fall when you drop them is intimately tied to the march of the planets across the heavens. And it took James Clerk Maxwell to explicitly state how the electrical phenomenon of lightning and a compass's unwavering ability to point north were just different manifestations of a single underlying phenomenon called electromagnetism. With the realization that electromagnetism was also the explanation of how light worked and governed all of chemistry, the beauty of Maxwell's equations became even more evident.

With this history of the unification of forces, it was natural to ask whether the three observed subatomic forces (strong, weak, and electromagnetic) were somehow related. The early 1960s brought a partial victory, when it was shown that electromagnetism and the weak force are really the same thing and that we should instead refer to them as the *electroweak force*. The first paper in the unification saga was written by Sheldon Glashow in 1961.

In order to unify the weak force with electromagnetism, Glashow theorized that four massless, force-carrying particles would govern the forces. These particles were hypothesized to be massless simply because that hypothesis is the easiest one and it worked. At the mathematical level, when a theory includes a mass term it is very difficult to work with and you sacrifice many convenient and simplifying mathematical properties. The most crucial point is Glashow's theory showed how the electromagnetic and weak forces could originate from a single and deeper phenomenon.

There was a significant complication, which was that the weak force didn't act like it was caused by the exchange of a massless particle. Massless and uncharged particles like the photon have infinite range. But massive particles have an extremely limited range, especially virtual massive particles, which can exist for only a very short time. When you think about it, if particles have a short range, this means it is hard for them to interact with their neighbors because often the neighbors are out of range. "Hard to react" is synonymous with "weak," and we can now understand the reason the weak force is weak. It is because the W and Z bosons, which are the particles that transmit the weak force, are heavy. However, we are getting ahead of ourselves. In 1961, the W and Z bosons hadn't been discovered yet. (They were discovered at the CERN Sp$\bar{p}$S in the early 1980s). Glashow and his contemporaries only knew that the weak force was weak.

Where does this leave us? Electromagnetism acts as if it is caused by the exchange of light particles (photons), while the weak force acts as if its cause involves heavy particles. Glashow's theory unifying the weak and electromagnetic forces required the force particles be massless. So we appeared to be at an impasse, or at least the theory unifying the two forces seemed to be false. In order to rescue this unification idea, we needed to find some mechanism that could be added to the theory and that gave mass to some particles and not to others. Because of some of the technical details of the underlying math, physicists call this mechanism *electroweak symmetry breaking*, or EWSB. Symmetry breaking means somehow taking objects that were the same (symmetry) and making them different (breaking), like giving mass to some particles but not others. But that is just insider jargon. For our purposes, the most crucial point is that, if we find the correct solution, we can both retain the idea that electromagnetism and the weak force are two components of the same thing and account for the difference in the mass of the particles that transmit the two forces.

Introducing Higgs and "His" Boson

This solution turned out to be proposing that there is an energy field in the universe. While this energy field is now universally called the Higgs field, Peter Higgs shouldn't get all the credit. (It is worth noting that Peter Higgs has long protested the name, but this has been a losing battle.) The year 1964 was when several groups independently proposed bits and pieces of the Higgs field. On June 26, François Englert and Robert Brout submitted their paper *Broken Symmetry and the Mass of Gauge Vector Mesons* (published August 31). Peter Higgs wrote a paper *Broken Symmetries, Massless Particles and Gauge Fields*, submitted on July 24 (published September 15). Gerald Guralnik, Carl Hagen, and Tom Kibble's *Global Conservation Laws and Massless Particles* was received on October 12 (published November 16). And Peter Higgs' paper *Broken Symmetries and the Mass of Gauge Bosons* was submitted on August 31 (published October 19). Luckily for writers—and readers—everywhere, we don't call this idea the Brout-Englert-Guralnik-Hagen-Higgs-Kibble mechanism, but this is a more accurate name. If you're interested in knowing more about the history and personalities, Ian Sample's book *Massive*, listed in the suggested reading, tells the tale in an in-depth and engaging way.

We will just use the use the common name of "Higgs mechanism" or just "Higgs," even though we can see that this does a significant disservice to at least five other exceptional and creative minds.

Higgs did have one insight that perhaps can explain why his name is associated with the energy field that might be the origin of the mass of subatomic particles. When he submitted his original paper (the second one mentioned above), it was initially rejected by the journal. (This is not an uncommon occurrence.) He revised the manuscript, and, among the revisions was the observation that, if the energy field he was proposing was real, it meant that a new particle existed. We now call this the Higgs boson.

The Higgs field was married to Glashow's unification in 1967 by Steven Weinberg and Abdus Salam. The original concepts of the Higgs field and the Higgs boson were written for abstract theories that weren't necessarily thought of as having a physical reality. It was Weinberg who adapted the Higgs idea to give mass to the W and Z bosons, but there are more subatomic particles than those. Weinberg's paper actually went further and described how the Higgs field could also be responsible for the masses of the fundamental fermions, such as quarks and leptons. (See chapter 2 if you need a refresher course on these particles.) It isn't mandatory that the Higgs field is the source of the mass of both the fundamental fermions and bosons—the latter consists of force-carrying particles (such as photons)—but it is certainly an economical hypothesis and is now a commonly held idea. Of course, as with all theoretical models, whether the Higgs field is the origin of boson mass, fermion mass, both, or neither is something that has to be settled empirically. The smart money is on "both," but even now this is not yet 100% assured. As this book goes to press, the picture is becoming clearer, but a definitive statement is still just a tiny bit premature.

It is also extremely important to note that the Higgs field as it was originally postulated didn't emerge naturally from some deeper theory. It was something invented just to provide a reason for there being a break in the symmetry of a theory and make sense out of why some particles had mass and others had none when some was expected. (We return to this concept of breaking symmetry below.) It's a Band-Aid really, not a serious theory. We will return to this observation when we discuss some of the questions raised if the Higgs field should turn out to be real. It is probably worth noting that subsequently there have been other theories proposed from which the Higgs mechanism appears, although none of those proposed theories are widely accepted.

For the moment, we don't need to worry about that. The first thing is to understand how the Higgs mechanism works. After all, the Higgs boson is probably the only subatomic particle that has made the lead story on CNN, BBC, the *New York Times*, hundreds of various television, radio, and print media, along with countless blogs. Just what *is* the Higgs field and the associated Higgs boson?

The "God Particle"

Before telling that story, it is perhaps worth talking about a common term used in headlines discussing the Higgs boson. This term is the *God particle,* which is also the name of a book by Nobel Laureate Leon Lederman and science writer

Dick Teresi. This book was first published in 1993 and covers several thousand years of the search for the ultimate building blocks of matter from Greek philosopher Democritus's atoms to modern-day quarks. However, the title of the book referred to the Higgs boson. According to Lederman, there were two reasons why he called it the God particle. The first is "[the particle is] so central to the state of physics today, so crucial to our understanding of the structure of matter, yet so elusive, that I have given it a nickname." That's the nobler reason. The second reason is because "the publisher wouldn't let us call it the Goddamn Particle, though that might be a more appropriate title, given its villainous nature and the expense it is causing." Now I know Leon, at least well enough that he penned a foreword for one of my books, and in addition to being an incredibly brilliant and accomplished man, he has a well-developed sense of humor. It's hard to really know if that second reason really happened or is what my Gaelic aunt used to call an Irish Truth—which is simply a story that is so good that it should be true. I'm not sure I really believe the second story, but it doesn't matter; it's a good story and I hope you repeat it. It's probably worth noting that Dick Teresi has been quoted as saying that he proposed the title as a lark and was shocked when the publisher accepted it. I don't know which tale is true, but Leon's story is better, so I'm going with it.

Whatever the antecedents, it is inevitable that some people will appropriate the name for their own purposes. Some point to the name and claim that this is proof that science is finally acknowledging the existence of God. Others take the opposite attitude and claim sacrilege. Physicists detest the term. In fact when Peter Higgs was asked about it, he said, "I really, really don't like it. It sends out all the wrong messages. It overstates the case. It makes us look arrogant. It's rubbish." He then added: "If you walked down the corridor here, poked your head into people's offices, and asked that question, you would likely be struck by flying books."

No matter your view on religion, whether you are deeply religious or an ardent atheist, the term God particle is simply stupid. It is misleading, and it trivializes the views of both communities. It is, as Higgs so pithily put it, rubbish.

Analogies

There have been several analogies used to describe the Higgs field and the Higgs boson, each with its various strengths and weaknesses. A few of them even have the notable distinction of having been awarded a bottle of champagne for their clarity. In 1993, William Waldegrave, who was then the UK science minister, issued a challenge to physicists to answer the questions "What is the Higgs boson, and why do we want to find it?" on one side of a single sheet of paper.

Face in a Crowd

Physicist David Miller took a political approach. He alluded to an ex-prime minister of the United Kingdom, which was obviously meant to mean Margaret Thatcher. However, as a scientist, I shall slightly modify the example and replace the politician with Peter Higgs. The modified analogy goes something like this.

Suppose there is a cocktail party of LHC physicists, with scientists spread evenly across the room. Suppose further that the door is on one side of the room, while the bar is on the other side. If I were to enter the room and decided that what was needed to start the evening off right was an aperitif, I would head across the room and peruse the barman's list of European beers. As I crossed the room, I would no doubt encounter a couple of people I knew and stop to shake hands and say hi, but my passage across the room would be mostly unimpeded. We can say that, as a particle, I would not have gained mass as I passed through the field of partygoers.

Now, let's consider if Peter Higgs were to enter the same room and have the same taste for a good stout. He, too, would enter the room and make a beeline for the beer, but his experience crossing the room would differ from mine. As soon as the crowd saw him, they would gather around him, trying to tell him of their most recent attempts to measure and characterize his eponymous boson. Between the questions, the handshakes, and the occasional back slapping, his passage across the room would be much slower. Essentially, Higgs would have gained inertia (and by extension mass) through his interaction with the crowd. This is an apt analogy of how a particle gains mass through its experience with the Higgs field and is shown in the left panels of figure 6.1.

This analogy explains the Higgs field and how particles gain mass but does not explain the Higgs boson. For that, we will let Higgs get his drink and retreat to a different room and have the crowd return to its original configuration—evenly dispersed across the room.

To understand how the Higgs boson appears, we need to know that the Higgs boson is simply a clump in the Higgs field. Using our crowd analogy, a small concentration of people in the otherwise evenly dispersed crowd is our analogy for the Higgs boson. To get an idea how such a clump might appear, we must inject a rumor into the room, say, that researchers at the Fermilab Tevatron had made a major discovery. This rumor enters the room by someone near the door telling a nearby member of the crowd. Since only a few people might have heard the rumor, and only poorly at that, other people in close proximity would collect near the first person to hear the rumor to find out what was said. This would lead to a local clumping of the crowd. People far away wouldn't realize that a rumor had

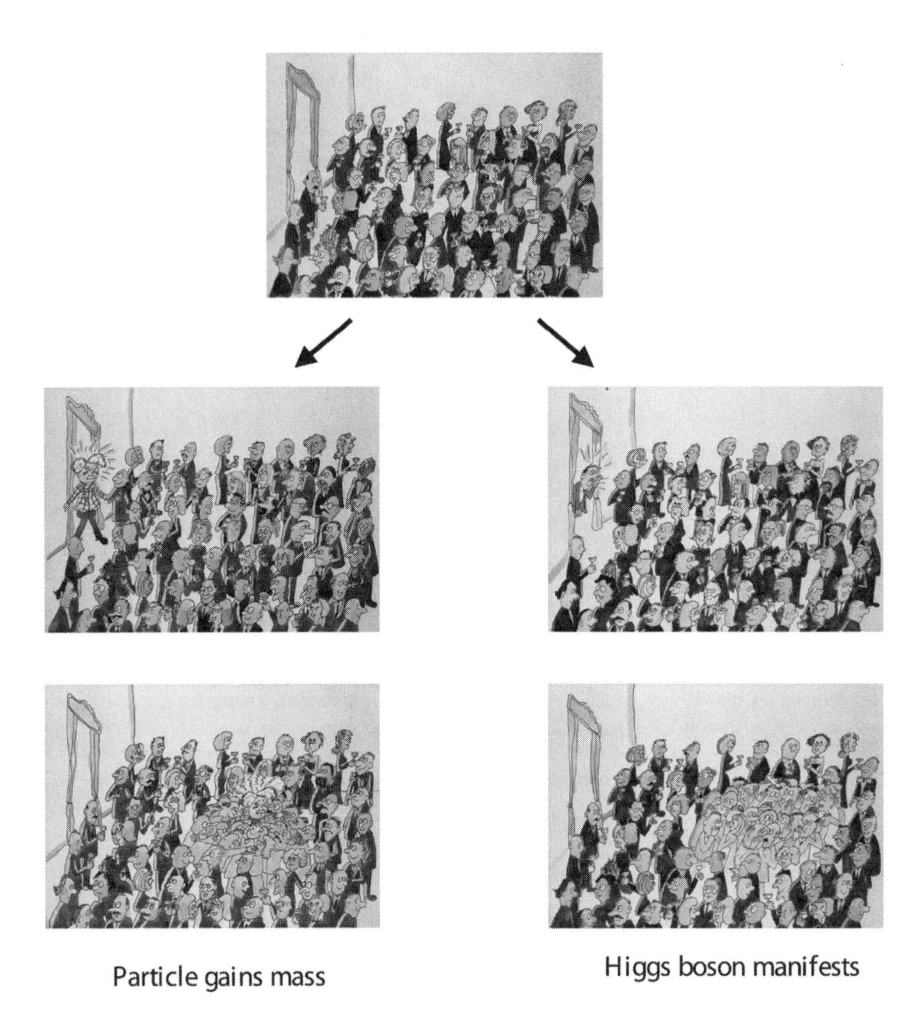

Figure 6.1. In an analogy in which a crowd stands in for the Higgs field, we see that a celebrity can have difficulty crossing the room (*left*), standing in for a particle effectively gaining mass as it passes through the Higgs field. A rumor can create a similar clump in the crowd that represents the Higgs field (*right*). The clump represents the Higgs boson. Figure courtesy of CERN.

entered the room, and they wouldn't move from their dispersed configuration. However, people who were close enough to the clump of people listening to the rumor would know something was going on, and they'd drift over to that group. Meanwhile, the first to hear the rumor would go back to their original positions, no doubt talking about what was said. By this process, the clump of people in the crowd would move across the room, with new people coming over to hear

Figure 6.2. The ease with which a barracuda and an aspiring sumo wrestler interact with the water can serve as an analogy for how the Higgs field gives mass to fundamental particles. Depending on the degree to which they interact with the field, they will have more or less mass. More streamlined objects—the barracuda in the image and a photon in the Higgs field—move more easily and have less mass. Heavier objects—the aspiring wrestler and a top quark—have more resistance to the field they are in and more mass. Figure courtesy of Dan Claes.

what the buzz was about and others drifting away, their curiosity satisfied. This analogy is illustrated in the right panels of figure 6.1.

A Fish in Water

Roger Cashmore took a different explanatory approach. He invoked the idea of viscosity and friction as a metaphor for mass. As I did with Miller, I shall embellish Cashmore's analogy to make it more dramatic. Let's start by imagining that we are in water. Water fills in for the role of the Higgs field. A barracuda, being supremely streamlined, can move through the water without difficulty. A physicist would say that the fish interacts only a little bit with the water, just as a photon ignores the Higgs field and remains massless. Continuing with our analogy, the barracuda can be considered a low mass particle.

In contrast, my buddy Eddy—aspiring sumo wrestler and no stranger to donuts—can only slog slowly through the water. A metaphorical top quark, Eddy interacts a lot with the medium. Figure 6.2 illustrates how these two stand-ins for subatomic particles behave in water.

Generalizing the analogy, we see that various immersed objects will interact more or less with the surrounding water and so resist motion to varying degrees depending on their mass and other factors. Similarly, the Higgs field imparts mass to particles, giving them a different amount of inertia.

While the water is analogous to the Higgs field, in the press what we hear is that what physicists are trying to find is the Higgs boson. How does our analogy handle this elusive particle? Well, we need to remember that, while water is a fluid and is the source of the viscosity that drives our analogy, we also know that water is made of countless individual molecules containing hydrogen and oxygen, specifically H_2O. Those molecules are metaphorical Higgs bosons. The two different aspects of water—liquid and countless molecules—correspond reasonably well to the two different appearances of the Higgs mechanism: the field and the boson.

Now, we should be cautious. This is an analogy, and all analogies are flawed. For instance, if you shoot a bullet into a pond, the bullet will eventually slow and stop. The constant interaction with the water makes the bullet lose energy and speed. However, we may recall that Newton's first law is "a body in motion tends to stay in motion." A massive particle, traveling in empty space, doesn't slow down and stop. Thus we have unveiled a limitation of the analogy. Still, the analogy has great intuitive power and thus, in spite of its limitations, I mention it here.

Wiggling Pencils

Tom Kibble explained the Higgs field as a table full of pencils balancing on their tips with wiggles in the pencil's balance filling in for the Higgs boson. His explanation focuses on the mathematical properties of the theory. This analogy is perhaps the most mathematically accurate, although it is not as intuitive as the earlier examples.

Recall that the Higgs mechanism was invented to explain the break in symmetry between the theoretical massless particles transmitting the electromagnetic and weak forces. This breaking of symmetry results in the photon and W and Z bosons that we have observed. When scientists talk about symmetry, they are thinking mathematically. If something is symmetric, it simply means that you take an action and nothing changes. For instance, take a clear crystal ball and rotate it. After the rotation, it looks just like it did before. Scientists and physicists would say that this action demonstrated the simplest form of symmetry. A real action resulted in no noticeable change.

So, Kibble's analogy focused on the symmetry of the balanced pencil. A photograph of the pencil's eraser, taken from above, could be rotated in any direction, and the photo would be essentially unchanged. In this sense, the orientation of the pencil is symmetric. However, were the pencil to fall over and lie on its side, the symmetry would be broken. Suppose when viewed from above, the pencil had fallen such that it appeared horizontal. If you took a photo of that pencil and rotated it 90°, the pencil would then appear to be vertical. The situation would no longer be symmetric.

While the symmetry is important, Kibble's analogy illustrated an important feature ignored by previous attempts. This is the mechanism of how the symmetry is broken. The Higgs field is said to "spontaneously" break the symmetry between the electromagnetic and weak forces. This means that it happens naturally; nothing out of the ordinary need be done. Kibble's analogy of balanced pencils is particularly apt on this point. A pencil, balanced on its tip, is symmetric when viewed from above. However, even if you do nothing, the pencil will naturally fall over and break the symmetry.

Like the blind men inspecting the proverbial elephant, each of these analogies illustrate a different facet of the nature of the Higgs field and its associated boson. None of these analogies are perfect and all of them fail when inspected closely enough. This perhaps is a good time to remind the reader of the dangers of believing any analogy. For researchers working on the LHC, it is a common occurrence to receive each month several long e-mails or bulging envelopes from amateur scientists—people with only a modest training in physics. Some of these communications are about a theory they have devised that is intended to make a large advance in our understanding of elementary physics. A separate class of communications is a thoughtful criticism of an analogy a person encountered that has been used to explain this theory or that. Often these criticisms are quite accurate and the author passionately believes that he or she has somehow invalidated the entire framework of (for example) the Higgs theory. However, frequently all these writers have done is expose a weakness in the analogy that is already well known to physicists.

The message is to accept the analogies for what they are: simple explanations that illustrate some particular feature of the more complex situation. Fully understanding the Higgs theory takes significant study and advanced mathematics. Scientists such as myself who take it upon themselves to try to explain the fascinating world of modern physics research do the best we can to find the most comprehensive analogy we can imagine, one that will explain as many features of the theory as possible. If you the reader devise an even better analogy, we'd all be happy to hear it. Be warned however. If I hear of a better analogy, I will immediately steal it. On this, I have no shame whatsoever. As the saying goes: lesser minds borrow, genius steals.

Slightly More Technical Descriptions

Modern quantum field theories treat everything in the quantum realm as a field. So, there is a photon field and an electron one. There are up quark fields and down quark ones. Pick a fundamental subatomic particle and it has a field. The particles are quantized vibrations of that field. To get an idea what that means, we must again turn to analogies.

Suppose we take a rope and tie it to the bottom of a post in the ground. We then lay the rope on the ground and stretch it out so it is nearly, but not quite, taut. We can call this rope the "bump field." Now, take the free end of the rope and give it a fast twitch. You'll see a bump appear in the rope and move away from you. One twitch means a single bump. Recall that the word "quantum" means "something that comes in a single unit." For instance, a person is the quantum of a crowd. Mix enough individuals and eventually you build up the crowd.

In the rope, you have a single bump moving in the bump field. This is a quantum bump. In the same way, the Higgs boson is a localized quantum vibration of the Higgs field (a bump in the Higgs field if you will) just as photons are quantized bumps of the electromagnetic field. When we search for Higgs bosons, what we're really doing is looking for those bumps in the field.

Our last foray into giving you a taste of the Higgs boson is a description adapted from a discussion by Flip Tanedo, a young theoretical physicist with considerable skill at explaining technical ideas in an approachable way.

Let's recap what we've learned. The known and observed force-carrying particles are the massless and neutral photon (electromagnetic carrier), the massive and neutral Z boson (weak force carrier), the massive and charged W^+ and W^- bosons (weak force carriers), and (now very likely) the massive and neutral Higgs boson. Whatever the theory says at the start, that's where we have to end up.

From here, we move into unfamiliar territory. The original electroweak unification theory required four bosons, all of them neutral. Physicists called these W_1, W_2, W_3, and B. You've never heard of them, because, like the flour, water, and yeast that make up bread, they cannot be seen in the final product but are essential in creating it. We could only observe them if the Higgs boson (or something like it) didn't exist. But the Higgs (or Higgs analog) does exist and this fact has erased the existence of these four particles. These bosons are introduced in figure 6.3.

Here's something else that will surprise you. The theory of the Higgs field and the associated boson doesn't predict just a single Higgs boson. In fact, it predicts that four exist! They are the H^+, H^-, H^0, and h. You'll see in a moment why you've never heard of these particles, but they are what the theory claims.

Figure 6.3. Before the Higgs boson's existence was confirmed, the electroweak unification theory predicted four massless particles that generate the weak and electromagnetic force: W_1, W_2, W_3, and B bosons. The B boson is shown off by itself in a different color because it is somewhat different than the W bosons, which are more closely related. Figure courtesy Flip Tanedo.

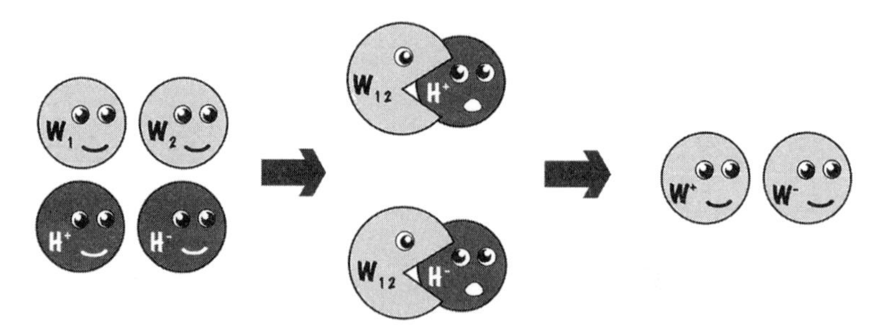

Figure 6.4. The W_1 and W_2 bosons gang up on the H^+ and H^- Higgs bosons, turning into the W^+ and W^- bosons. Figure courtesy Flip Tanedo.

(I should caution you here that in chapter 7 we will encounter a theory that posits yet another set of charged Higgs bosons. Those charged Higgs bosons and these charged Higgs bosons are completely different!)

When the Higgs theory encounters the electroweak theory, the result is the particles we actually observe. In this scenario, what happens is that the W_1 and W_2 combine in one way and gang up on the H^+ particles, resulting in the W^+ particle. Similarly, the W_1 and W_2 particles gang up in a different way and combine with the H^- particles to make the W^- particle. This process is illustrated in Figure 6.4.

In addition, the W_3 particle and the B particle combine in one configuration, resulting in the massless photon (γ). Those two particles (the W_3 and B) can also combine with the H^0 particle to make the Z boson. This process is visually represented in figure 6.5. So that's how we get the four observed electroweak bosons (W^+, W^-, Z, γ).

Where does the Higgs boson fit into all of this? Well, you'll recall that the Higgs theory predicted four particles (H^+, H^-, H^o, and h) and that we haven't used the h particle anywhere. That h is the Higgs boson that you hear about in the press. Thus, when you think about it, even before the recent hullabaloo

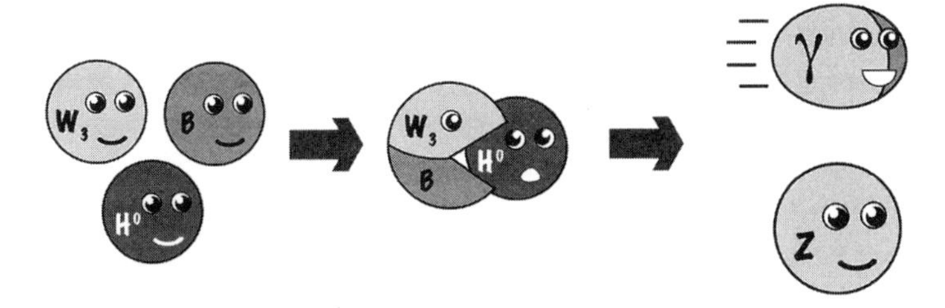

Figure 6.5. The W_3 particle and the B particle gang up on the H^0 particle to make the Z boson. The W_3 particle and the B particle can also combine in a different way resulting in a photon. However, they cannot combine with the last of the Higgs bosons, the h particle. That means that the photon doesn't gain mass. Figure courtesy Flip Tanedo.

about the Higgs boson, we discovered the presence of three of the four Higgs particles when the Z and W bosons were observed in the early 1980s. On the one hand, the Higgs boson buzz in July of 2012 was just tidying up a very important loose end. On the other hand, if we hadn't tugged at that loose end and found the Higgs boson, we could have rapidly unraveled the entire theory.

Making a God Particle (see how ridiculous that sounds?)

Thus far everything we've spoken of has been rather theoretical. What we need to do is to turn our attention to the experimental side of things. After all, the way we are able to validate this whole crazy Higgs idea is by finding the boson Peter Higgs predicted in the fall of 1964.

It is important to think for a moment about what a discovery means in particle physics (or any science, for that matter). There really are two different ways in which one can make a discovery. The first is when you see something completely unexpected, like if you were walking down the street and exclaimed, "Holy cow! That's a leprechaun!" Assuming that you're not crazy, that is a true discovery. The second version is when there is a clear prediction from some theoretical idea and you investigate to see if the idea is true. An example of this kind of discovery might be like when someone suggests that wearing a coat and tie might get you better service when you're in the airport. You could try the experiment and see if the proposal was in fact correct.

Searching for the Higgs boson was an endeavor of the second type. The Higgs theory is completely spelled out. The manner in which the Higgs boson is thought to be both to be created and to decay is completely predicted. You, as an experimenter simply test to see whether the predictions are correct. This, of course, does imply that you could miss something interesting by focusing exclusively on these specific theoretical predictions, thus scientists do look at lots of things (i.e., try to make discoveries of the first kind). However, strict

Table 6.1 Higgs bosons are produced at the LHC in various ways.

Process name	Debris in addition to Higgs boson	Relative frequency
Top/antitop fusion	1 top quark & 1 top antiquark	1
W/Z bremsstrahlung	W or Z boson	12
WW or ZZ fusion	Two quarks	20
Gluon fusion	None	250

Note: This is a listing of some of the more common ways. Realize that "more common" still means incredibly rare. Note further that the relative frequencies listed here are for a specific mass of the Higgs boson (125 GeV).

Higgs hunters had a specific program laid out before them. As they searched for Higgs bosons of different masses, they used the predicted production and decay modes to guide their search strategies.

Let's look at how Higgs bosons are made in an accelerator. At the LHC, pairs of protons are collided together at high energy: 7 TeV (2011), 8 TeV (2012), and 12.5 to 14 TeV (2015+). In these interactions, a fraction of the collision energy is converted to mass under the auspices of Einstein's $E = mc^2$. So, how is that done in detail?

It turns out that there are many different ways to make a Higgs boson, although not all of them are equally likely. For flavor, I will mention four and provide some information about them in table 6.1 and a representation of them in figure 6.6. Remember that at the LHC all collisions occur between two protons, although the real action is between particles found inside the protons.

1. *Top/antitop fusion*. A rare kind of collision that occurs is what physicists call "Higgs boson production with associated top quarks." In each of the protons, a gluon—a particle associated with the strong force that we met in chapter 2—briefly splits into a top/antitop quark pair. A top quark from one of the protons fuses with an antitop quark from the other proton and makes a Higgs boson. Thus we look for the signature for this kind of collision: the decay debris of the Higgs boson, plus the debris of the uncombined top and antitop quarks. This particular interaction is pretty rare, but we will use it as our standard of rareness. For our purposes, we will call the interaction probability for top/antitop fusion to be 1 unit of probability, and the probability of all other interactions will be given as a multiple of this rare occurrence.

2. *W/Z bremsstrahlung*. The next most probable interaction is called *W/Z bremsstrahlung,* from a German term that means "braking radiation." The classical definition of bremsstrahlung is when an electron slows down in material and emits a photon. In this instance, a W or Z boson kind of "slows down" and emits a Higgs boson. A more accurate description is that a light quark from one proton combines with an antiquark from the other proton in a collision and makes a

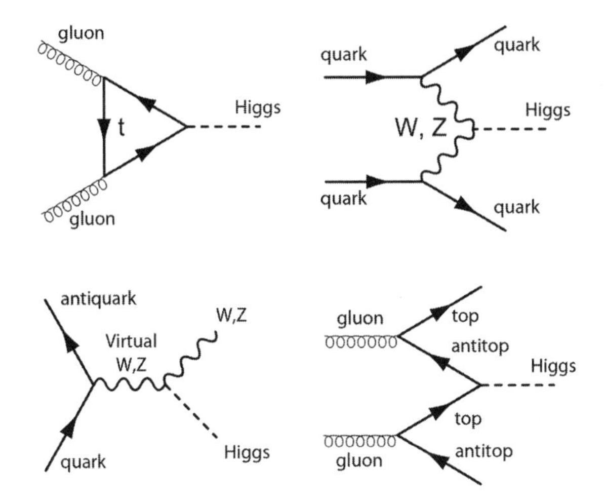

Figure 6.6. This figure illustrates how the Higgs boson is frequently made. While sometimes it is made in isolation, there are several instances in which the Higgs boson is made with a companion particle or particles. *Top left*, gluon fusion; *top right*, WW/ZZ fusion, *bottom left*, bremsstrahlung, *bottom right*, top/antitop fusion. t = top quark.

"virtual" W or Z boson. "Virtual," in this context, means that the mass of the W or Z boson is not "right" and is different from the values given in chapter 2. This counterintuitive situation is made possible by the Heisenberg uncertainty principle, which allows particles to briefly exist with the wrong mass. This virtual W or Z boson then decays into a Higgs boson and a "real" (i.e., with the right mass) W or Z boson. The signature of this kind of interaction is the decay debris of the Higgs boson and the W or Z boson that was produced along with it. This interaction occurs twelve times more often than the top/antitop fusion process.

3. *WW or ZZ fusion.* This is the next most common interaction. It is also known as *vector boson fusion* (as the W and Z bosons are called vector bosons for reasons beyond the scope of this book). In this process, both protons emit quarks, which subsequently both emit either a W boson or a Z boson. The two W bosons or the two Z bosons fuse and make a Higgs boson. Like in the W/Z bremsstrahlung case, the W and Z bosons are virtual (the "wrong" mass), but the Higgs is not. The signature of this production process is that we see the debris of the Higgs boson, as well as the debris of the two quarks after they emit the W or Z bosons. This interaction is twenty times more likely than top/antitop fusion.

4. *Gluon fusion.* This is the most common way in which a Higgs boson created. In this case, both protons emit a gluon, each of which emits a top/antitop quark pair. This is a little like the top/antitop fusion, except the top quarks are

emitted in a "loop." What really happens is one gluon turns into a top/antitop quark pair and the second gluon knocks the quark/antiquark pair back together to form a Higgs boson. Since quantum mechanics doesn't allow you to know which gluon was the emitter and which knocked the quark/antiquark pair back together, it is drawn in the manner seen in figure 6.6. This most common process occurs 250 times more often than the baseline top/antitop fusion. The signature here is simply the decay products of the Higgs boson. Table 6.1 summarizes the probabilities and figure 6.6 illustrates what is going on.

From our discussion here, we see that gluon fusion is quite common. Surely we focus on this case, correct? While it is certainly true that we pay attention to this particular production mode, the fact that only the Higgs boson is made (with its decay products) means that it is easier for far more common interactions to fake a Higgs boson. We will discuss the decay modes in a moment, but the most common decay mode of a low-mass Higgs boson (as hypothesized in the Higgs theory) is into a bottom quark/antiquark pair. Making bottom quark/antiquark pairs by means of far more pedestrian channels is millions and billions of times more probable than having them come from the decay of a Higgs boson. So, seeing them could create a sort of a false positive for seeing a Higgs boson. Thus, it is sometimes more productive to use the rare kinds of collisions to identify events in which Higgs bosons are actually made and reject look-alikes. And, of course, since the Higgs theory predicts all of these ways in which the Higgs boson can be made, to be sure we have found the Higgs boson we will have to eventually see it in all these production modes (and more!) and also to verify that they are created at the predicted rates.

To drive home the idea of the difficulty in identifying events in which the Higgs boson was created, let us consider a comparable situation involving diamonds and cubic zirconia, which look enough like diamonds to fool most people. If you had a bucket full of cubic zirconia in which one diamond was placed, you would be hard-pressed to identify the diamond. Finding Higgs bosons decaying into bottom quark/antiquark pairs while awash with pairs from more ordinary processes is analogous, except to be more accurate the "pail" would be a cube perhaps 10 feet (3 meters) on a side.

It is perhaps important to remind the reader that the numbers discussed in the text and mentioned in table 6.1 are actually dependent on the mass of the Higgs boson. Before the discovery of the Higgs boson, we didn't even know if the Higgs boson existed or not. While we now know more, at the time, we didn't know its mass. Thus we had to adjust our expectations and search at all possible masses. The whole situation is quite tricky. I will not dwell on how the production modes depend on the mass of the Higgs boson but rather will give a sense of the range of the allowed variation in the discussion on how it is predicted to decay. However, you should remember that the complexity discussed there applies here as well.

Decay

It is said that the only things that are inevitable are death and taxes. While I expect that Higgs bosons probably can duck taxes, but even they are hounded by death or, more accurately, decay. In the subatomic world, anything that can happen eventually will and often extremely quickly. While the lifetime of a Higgs boson depends on its mass, it is exceedingly short. To give a sense of scale, the beam pipe at the LHC has a radius of about an inch. In order for a particle to live long enough to hit the beam pipe, it has to live about a tenth of a billionth of a second. While this does seem incredibly short, Higgs bosons live far shorter than that, less than a trillionth of that fleeting time.

Because of this, the only way we will ever "see" a Higgs boson will be to see its decay products, which physicists call *daughter particles*. At the outset, we don't know if the Higgs boson exists or its mass if it does exist. However, if it does exist and we know the mass, the Higgs theory completely describes how often it can decay and in which way. So, as we look for the Higgs boson at various masses, we simply change our expectations as we go. This is the approach that was employed in the search for the Higgs boson.

You will recall that the Higgs mechanism is supposed to be the thing that gives fundamental particles their mass. One consequence of this is that particles that interact more strongly with the Higgs boson have more mass. Accordingly, the Higgs boson decays preferentially into the heaviest particles it can.

So, if we look at table 2.1, we can find the list of known particles and identify those that are most massive. They are (in descending order of mass): top quarks (173 GeV), Z bosons (91 GeV), W bosons (80 GeV), and bottom quarks (4.5 GeV; where GeV is a measure of energy and approximately 1 GeV is the mass of a proton). Because you can't get something for nothing, the Higgs boson's mass must be at least twice the mass of the objects into which it decays. For instance, in order to make two 173 GeV top quarks, you need a minimum of 346 GeV to start.

If its mass is above 182 GeV, it can decay into pairs of Z^o bosons. Above 160 GeV, the daughter particles can be a $W^+ W^-$ pair, while between 9 and 160 GeV, the way to look for Higgs bosons is to try to find bottom quark/antiquark pairs.

When one looks very carefully at the theoretical predictions, one sees that the situation is slightly more complex, as illustrated in figure 6.7. The Higgs bosons definitely can decay in the ways described in the previous paragraph. But subtle physics effects also come into play that give an edge to W and Z bosons in the competition as to what decays will dominate. The net effect is that below a

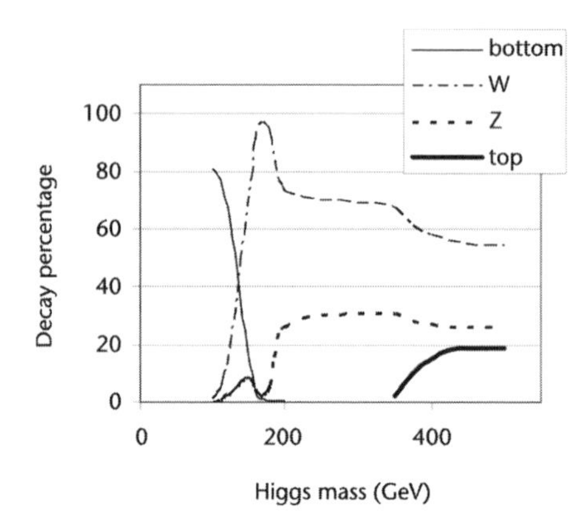

Figure 6.7. Decay percentages of the Higgs boson. If the Higgs boson mass is low, it preferentially decays into bottom/antibottom quarks, although, at higher masses, pairs of W bosons are preferred.

possible mass of the Higgs boson of 135 GeV, it will decay predominantly to bottom/anti-bottom quark pairs, while above that the preferred decay mode is into W boson pairs. However, the other decay modes we've listed will be possible, in addition to ones not mentioned here. LHC physicists looked for and continue to look for all of them.

Given that bottom quarks, W and Z bosons, and top quarks are the most likely ways in which the Higgs boson will decay, we need to explore how these decay particles will themselves be observed in a detector. After all, we never see the Higgs boson itself, but only infer its existence from its daughter particles. The problem is that all of these daughter particles decay as well. Top quarks decay 100% of the time into bottom quarks and W bosons. This gets us to the final point. Any LHC detector that wants to look for the Higgs boson predicted by current theory had better be able to measure well W and Z bosons and bottom quarks. So let's briefly talk about how Z bosons, W bosons, and bottom quarks decay.

Because of the nature of the strong force, quarks don't like to be alone. If a quark is pulled away from other quarks, the strong force acts a bit like a glob of water thrown from a glass. First there is a slug of water, but surface tension pulls it apart into individual water droplets. Figure 6.8 illustrates this analogy.

Thus a single quark will turn into many particles, all traveling in the same direction. This stream of particles is called a *jet* and resembles the blast of pellets that come out of a shotgun. Just as a single shotgun cartridge turns into many pellets, all going in generally the same direction after they leave the barrel, a single quark turns into many particles all traveling in the same direction

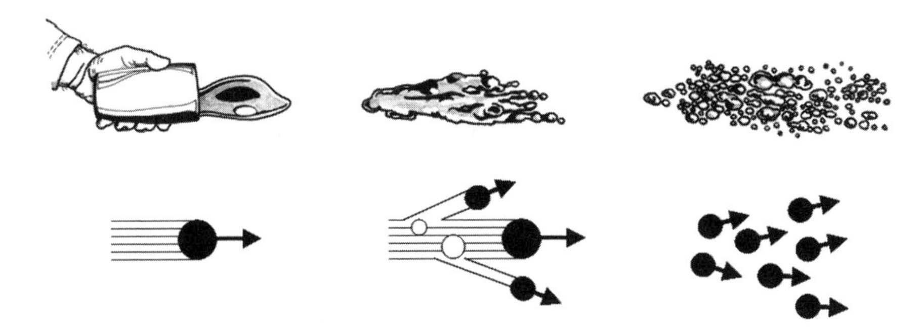

Figure 6.8. Just like a slug of water will turn into water droplets, a single quark will turn into many particles. The physical mechanism is quite different, but the essentials of the process are quite similar. Water drawings courtesy of Dan Claes.

after they leave the vicinity of other quarks. Obviously, the physical mechanisms governing the two phenomena are very different, but the mental picture is valuable even so.

All quarks turn into jets after a particle collision except for top quarks, which turn into a bottom quark and a W boson before there is time for a jet to form. But this daughter bottom quark does form a jet.

The Z and W bosons can also decay into quarks. But these bosons have an option not available to quarks. The bosons can also decay into pairs of leptons, of which the electron is the most familiar. Figure 6.9 shows the various ways that the Z and W bosons can decay. From an experimenter's point of view, the most interesting decays are those involving electrons, muons, and neutrinos. These particles are quite distinct and can more easily identify that a W or Z boson was created in the collision.

Let's combine what we've learned and imagine what went through the mind of experimenters before the discovery of the Higgs boson. Suppose the Higgs boson was very light, say 115 GeV (or about 115 times heavier than a proton). We see in figure 6.7 that the Higgs boson will most frequently decay into a bottom quark and antimatter bottom quark pair. Thus to see a light Higgs boson, you need a detector that can measure two jets very well, with both coming from bottom quarks.

If the Higgs boson is much heavier, say 160 GeV, then the Higgs boson most likely decays into a W^+W^- pair. Although quarks from W boson decay can also produce jets, jets are also made by more ordinary quarks just getting knocked out of the proton (and jets from ordinary quarks are much, much more common). Thus, experimenters are more interested in the decays involving leptons, because they are easier to identify as having originated in the decay of a

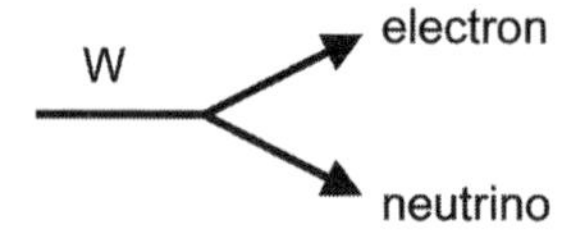

W decay modes	
Decay mode	Frequency
W → quark + antiquark	67%
W → electron + neutrino	11%
W → muon + neutrino	11%
W → tau + neutrino	11%

Z decay modes	
Decay mode	Frequency
Z → quark + antiquark	70%
Z → neutrino + antineutrino	20%
Z → electron + positron	3%
Z → muon + antimuon	3%
Z → tau + antitau	3%

Figure 6.9. The top drawing is a representative decay of a W boson, while the tables show the frequency by which the W and Z bosons can decay. The missing 1% in the right hand table is a round-off error.

W boson. Say one W boson decays into an electron and neutrino and the other decays into a muon and neutrino. Since both the Higgs and W bosons decay rapidly, what we as experimenters would see is an electron, a muon, and two neutrinos. Since neutrinos don't interact very much with matter, they escape undetected. So we would see an electron, a muon, and missing energy.

Missing energy makes things a little tricky. Consequently, even though the Higgs boson decays into W boson pairs about twice as often as it does into Z boson pairs, the fact that Z bosons decay into pairs of charged leptons make it a very attractive way to look for Higgs bosons. For instance, if a Higgs boson decays into two Z bosons and both Z bosons decay into electron/antimatter-electron pairs, the experimental signature of a Higgs boson would be the two electrons and two antimatter electrons (positrons) in your detector. If one of the Z bosons decayed into a muon/antimatter-muon pair instead, then you'd see a pair of electrons and a pair of muons. The most important point is that, when a Z boson decays, you see both decay particles, while you don't for the W boson. The chain of Higgs to two Z bosons and then to four leptons has been called the "golden Higgs decay mode." You get an idea of just how powerful and clean this decay channel is when you realize that two Z bosons decay into four leptons only 0.4% of the time. This is a tiny, tiny fraction of Higgs boson decays, and yet experimenters still viewed this decay configuration as the way to most easily identify events in which Higgs bosons are created. Figure 6.10 shows some of the ways that Higgs can decay.

While I've emphasized the way that the Higgs boson prefers to decay into

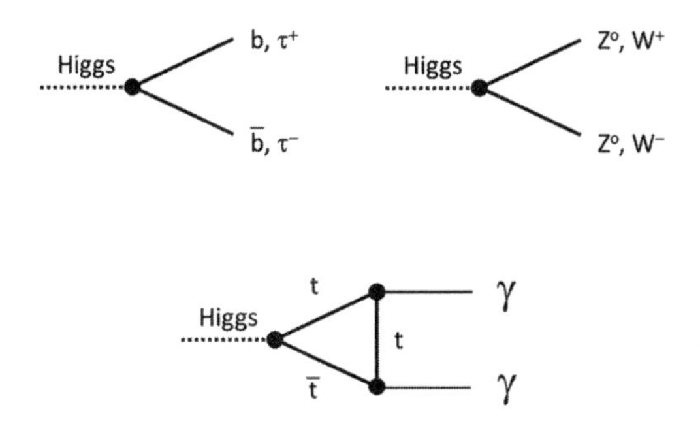

Figure 6.10. Higgs bosons can decay in a number of ways, for instance into pairs of heavy fermions, like antimatter and matter bottom quarks or tau leptons (*top left*) or pairs of heavy bosons, like Z bosons or W bosons (*top right*). However, one important decay is rare. The Higgs boson decays into a top quark/antiquark pair that annihilates and emits two photons.

heavy particles, there is a rather counterintuitive, but incredibly powerful, decay mode that needs to be considered.

The massless photon is the least likely particle into which you'd think Higgs bosons can decay. After all, interacting with a Higgs boson gives the particles their mass and, by definition, a massless particle cannot directly interact with the Higgs boson. Ah, but I've added a weasel word "directly." What does that mean?

It turns out that it is possible for the Higgs boson to decay into particles that then subsequently emit photons. This is an indirect decay and is therefore quite rare. In fact, it happens only 0.2% of the time. However, that 0.2% was thought to be one of the best ways in which to find the Higgs boson and, as we shall see, it played a pivotal role in the excitement of July 2012. The reasons for this are twofold: (1) making two high-energy photons by accident is pretty rare and (2) detecting photons and accurately measuring them is pretty straightforward compared with the more common decay modes.

Given the role of this counterintuitive decay mode in the discovery of the Higgs, it is worth exploring in more detail. Because the Higgs boson prefers to interact with massive particles—indeed because when it interacts more with certain particles it gives them more mass—if it can, it will interact with top quarks. Two top quarks have a combined mass of 350 GeV, which is higher than the mass the Higgs boson is expected to have. However, again quantum mechanics helps us. If the Higgs boson had a mass of (say) 150 GeV, it would be possible to decay into two top quarks each with the "wrong" mass of 75 GeV. As long as the top quarks didn't live too long, Heisenberg's uncertainty principle blesses this seemingly crazy occurrence. And, because top quarks have electrical charge, they can emit photons.

So, this is what happens. In sort of a backward version of the gluon fusion production mechanism, a Higgs boson decays into two virtual (i.e., wrong mass) top quarks, each of which emits a photon. The top quarks then annihilate one another to tidy up the whole situation. Thus, an event in which two high-energy photons are emitted is an interesting possibility. We'll see just how interesting in a moment.

The Hunt

The search for the Higgs boson has been a long one, beginning shortly after the particle was proposed in 1964. I will dispense with much of the prehistory of this storied search and simply mention the last big push prior to the start of LHC operations. Before the LHC was constructed, another particle accelerator inhabited the same tunnel. This accelerator was called LEP (for Large Electron Positron), and it supplied electron and positron beams to four superb detectors. Together, these detectors looked for Higgs bosons. While the tally of LEP physics successes is extraordinary, the Higgs boson wasn't among them. After about a decade of exquisite measurements, the LEP accelerator shut down for the last time in 2000. After a bit of excitement in its last days of running, in which the Higgs seemed to be playing peekaboo with the experimenters, they finally concluded that the Higgs boson wasn't going to be added to their list of honors. However, even though they didn't find any, the LEP experimenters knew that they would have found the particle if its mass was below the precise number of 114.4 GeV. At least they told subsequent searchers that looking for Higgs bosons with a mass under that was a waste of time.

On the top end of the mass range, there were theoretical reasons to reject a range of possible masses for the Higgs boson. While the reasons can be quite complex, they can be simplified to a single strong bit of circular logic wrapped up in mass. Recall that the Higgs field interacts with particles and gives them their mass. This is true for the Higgs boson itself. If the mass of the Higgs boson were high, this would mean that the Higgs field would interact with Higgs bosons quite strongly. This self-interaction has ugly and dramatic consequences, none of which were observed. One such dramatic consequence would be that the chance heavy particles would decay would exceed 100%. Thus, while theoretical reasoning isn't quite as compelling as a direct search, it is plausible to conclude that the mass of the Higgs boson wasn't huge, with a maximum of about 185 GeV being a possible limit.

As of the fall of 2008—the moment when the first beam was injected into the LHC—we knew that if the Higgs boson existed that it would have a mass in the range of 114 to 185 GeV. In an earlier book, I stated that the most likely spot is about 125 GeV or so. As we shall see, this turned out to be a remarkably prescient prediction, one that has led more than one of my collaborators to ask if I had any hunches for the mass of other undiscovered phenomena. Between you and me, it was just a lucky guess. But don't tell them. They still buy me drinks over the prediction.

The LHC turned on in the fall of 2008, only to be spectacularly damaged just a few days later in a confluence of insufficiently careful engineering, assembly problems, and just plain bad luck. The repairs took a little over a year and in the spring of 2010, the LHC came back to life. It was show time.

When the LHC turned on, it wasn't the only game in town. The Fermilab Tevatron had begun operations in 1987 and been running off and on for two decades. The Tevatron was a lower energy machine—2 TeV, compared with the LHC's design energy of 14 TeV—and collided beams of protons and antiprotons. In addition to the lower energy, the collision rate was also lower. Still, the two big experiments at the Tevatron—CDF and DZero—had been working a long time on the Higgs and, even though the LHC was obviously the new phenom in town, the old veterans weren't going to go down without a fight.

I was working both on the Fermilab DZero experiment and on the CERN CMS one in those days, giving me an insider's view of both groups. (This wasn't so unusual, as this was true of many of my colleagues.) Both groups were tense but professional. While everyone involved wanted to be the one to discover the Higgs boson if it existed, the most important thing was to get it right. False alarms and retracted discoveries do nobody any good.

The first to have something to say was the Tevatron. In July of 2010, the two Fermilab experiments combined their data, effectively doubling the amount of beam collected. It's a lot easier to say "combined their data" than actually do it. Some effects were totally independent between the two experiments (like their respective equipment), while others were exactly the same (like how they measured the amount of beam delivered by the same accelerator). Figuring out just how to combine the two group's measurements required a subtle understanding of . . . well . . . just about everything. Oh yeah, and it took some rather serious statistical fancy footwork as well.

After the usual careful work, critical evaluation, and occasional acrimonious shouting matches over the finer points of this and that, the thousand or so experimenters finally agreed. They could rule out the possibility that the Higgs boson had a mass between 158 and 175 GeV limits. If the Higgs boson existed, it was now constrained to be within 114.4 to 158 GeV or 175 to 185 GeV. In science, knowing where something *isn't* tells you to stop looking there and redouble your efforts where it might be found.

This announcement didn't cause incredible anxiety among the LHC experimenters; after all, they had just turned on a few months earlier and they were still in shakedown mode. But it did remind them not to take anything for granted. The future was theirs to have. But not if they didn't hustle.

The next big event was a year later, in July of 2011. In addition to the lower energy and lower beam brightness, the Tevatron experiments had another limitation that isn't obvious. They had been running for a decade. Now this doesn't seem like it should be a bad thing; after all, it means that they understood their equipment very well and all the bugs had been shaken out. But, in having taken data for a decade already, the year between the two Julys would only add 10% more data if everything were equal. The actual amount of data added was higher than that, due to the fact that the early years of Tevatron running were at lower beam brightness and further the venerable Tevatron was simply purring and cranking out the best beam it ever accomplished. But the principle held. In contrast, the beam recorded by the LHC in 2010 was really pretty pitiful. It was enough data to make fantastic measurements of known phenomena but really wasn't enough to really weigh in on the Higgs hunt. However, if 2010 was a trickle, then 2011 was a faucet cracked wide open. The beam recorded by the LHC in 2011 simply crushed the amount recorded in 2010. The LHC experiment's 2011 data by July were a little over thirty times larger than all the data recorded in 2010. (And if the 2011 rates were a faucet run wide open, then 2012's rates were a fire hose, but let's not get ahead of ourselves here.)

Using the data recorded by July 2011, the LHC experiments announced that they were able to rule out an even larger range of masses than the Tevatron did. ATLAS ruled out the possibility that the Higgs boson was in the range of 155 to 190 and 295 to 450 GeV, while CMS ruled out from 149 to 206 and 300 to 440 GeV. You'll note that the upper end of these numbers is higher than the 185 GeV already ruled out. The difference was that these were direct experimental limits and therefore even more persuasive. Direct limits are when you just use brute force to rule out a value. In indirect limits, you use a bunch of different measurements and combine them to infer something about a variable that you didn't measure directly.

That was when the Higgs baton was passed from the United States to Europe. It was a sad day for those of us working on the Tevatron experiments but, then again, many of us were the same people working on the LHC experiments. So, after a period in which we congratulated ourselves a bit and consoled ourselves a little more, we put aside such schizophrenic silliness and turned back to the hunt.

By November 2011, the LHC had collected considerably more data, and researchers at the ATLAS and CMS experiments announced at a conference that they had ruled out a greater range of possible Higgs masses. Now both experiments were giving similar limits. They confirmed the earlier LEP limit and could rule out masses below 114 GeV, but they could also now rule out masses in the range of 141 to 476 GeV. Ruling out isn't finding, and the race continued.

Just a month after the November announcement, the LHC experiments completed their preliminary analysis of the entire set of 2011 data (five times what was reported in July) and, on December 12, 2011, they had another update. Still no discovery, but the range of possible masses was even smaller, with the Higgs boson only still possible in the range of 115 to 131 GeV (ATLAS) and 115 to 127 GeV (CMS). By narrowing the range of possible masses, scientists knew where to redouble their efforts. Even more interesting, there were tantalizing hints as to the value of mass the Higgs boson might have. The data were by no means good enough to make a definitive statement, but it was certainly pointing at a mass of 125 GeV. However in December 2011, it remained a real possibility that subsequent analysis would rule out those small remaining ranges. This would mean that the Higgs boson didn't exist at all.

The LHC shut down for the annual winter break. This is usually is mid-December through mid-February. When it turned on in 2012, the LHC came up with a vengeance. The energy had been raised from 7 TeV to 8 TeV, and the collision rate was four or five times higher than in 2011. This looked to be the year. In March of 2012, the Tevatron experiments expanded the range of masses they could exclude, now excluding 147 to 179 GeV. But this addition wasn't competitive and cemented the fact that the LHC was now firmly in charge of the Higgs hunt. The Tevatron did see tantalizing hints of something new in the range of 115 to 135 GeV, but this was comparatively imprecise. Still, it was good to see a common message from two separate accelerators.

So, now we get to July of 2012. The LHC searches were running flat out. The amount of data accumulated from March to July in 2012 already was the same as recorded in all of 2011. It was time for an update on what the data had revealed.

The problem with doing a 2012 analysis is that the 2011 data pointed in a direction. The 2012 analyzers could potentially be biased, and this is simply not allowed. The point, after all, of collecting more data is to be utterly impartial in how the analysis is approached. Only then can you be assured that the new data support the original analysis, because that's what the data say and not because you made it happen (perhaps even unintentionally).

To guard against unwitting bias, ATLAS and CMS both blinded themselves from each other and further blinded their 2012 analysis from the 2011 data. "Blinding" in particle physics means you are not allowed to look at the interesting region of your data. You can look at data to verify that you can see known things, for instance, in regions where you know there is no signal. You can look for other known phenomena. But you can't look where the signal is thought to exist. Thus the 2012 analyses were not allowed to look at data anywhere near the mass region hinted at by the December 2011 announcement.

There was a little drama going into the first week of July. There was a conference scheduled in Australia called the International Conference on High Energy Physics, or ICHEP for short. It was running from July 4 to 11 in Melbourne and was one of the big particle physics conferences that year. Long before the start of the conference, physicists signed up to go because . . . hey, it's Australia . . . and because it was expected to be a great place to get updated on the 2012 physics results, although the Higgs hunt wasn't really expected to be a big player. In March and April of 2012, it was generally conceded that the LHC wouldn't have a definitive statement on the whole Higgs thing until later in the summer, perhaps not even until the Hadron Collider Physics (HCP) conference in Kyoto that November.

However people hadn't really banked on the quality of the beam delivered by the LHC and the almost-manic efforts of the Higgs boson analyzers. In the last week of June, the analyses unblinded themselves from the interesting data and found that a very strong (and confirming) Higgs boson signal was there.

A Statistical Interlude

Before we delve into the end game of the July announcement, it is worth spending a little time talking about statistics and the role they play in discovery in general and particle physics discoveries in particular. Here's a secret that isn't fully appreciated by most people: rare stuff happens.

You could, in principle, roll a single die ten times and have it come up sixes every time. You could, in principle, win the lottery twice in a row. You could flip a coin and have it land on its edge. None of these are likely, but all are possible. Statistics tries to help you figure out when something is a bizarre fluke that doesn't mean anything and when a weird observation is telling us that we've found something.

We'll return to particles in a moment, but it is easier to see the important ideas using a more familiar example. For example, let's talk about that die I mentioned earlier. Suppose someone thinks the die is loaded and that somebody has messed with it to make it come up with a six more often than the expected one-sixth of the time. What does statistics tell us about that? Well, what it tells us is the probability that all possible things can occur, even the rare ones. Let's take a specific example and roll this die twelve times. I show in table 6.2 what we expect purely by random chance.

The first thing we see is that the obvious answer is wrong. This obvious answer is that if we do something twelve times, each of which has a one-sixth

Table 6.2. Percentage of the time that a single and fair die, rolled 12 times (or 12 fair dice, rolled one time), will come up with the specified number of sixes

Number of sixes	Percentage of time expected
0	11.2%
1	26.9%
2	29.6%
3	19.7%
4	8.88%
5	2.84%
6	0.663%
7	0.114%
8	0.0142%
9	0.00126%
10	0.0000758%
11	0.00000276%
12	0.00000005%

chance of happening, then we should see this happen twice. But the chance of two sixes is 29.6%, which means that there is over a 70% chance that we'll see something other than two sixes. Further, we see that about one time in a thousand, we expect to see seven sixes and that even rolling a six twelve times is possible.

And this is a fundamental point. In experimental science, nothing is guaranteed. Everything is probability. So some guidelines have been set up. Note that these really are just guidelines. They're reasonable, but guidelines they are. The language of these guidelines comes from bell curves, normal distributions, or gaussians (these are three ways to say the same thing), but in the end it's really just about probabilities. The word *sigma* in a statistical context is a measure of the width of a bell curve. It comes from the Greek letter equivalent to S and refers to the standard deviation, or change, from the average. If you start from the middle of the bell curve and go "1 sigma" in either direction, this region will encompass 68% of the area of the bell. If you go "2 sigmas" in either direction, it will encompass 95% of the area, and so on.

That's all we're going to say about sigmas. We're actually really interested in the probabilities. When you read that a measurement is a "3 sigma" one, that means that what was observed could happen one time in 740 purely by accident. A 4 sigma measurement can occur by accident one time in 31,574, and a 5 sigma measurement can occur by chance about one time in 3.5 million. In particle physics, we have a semi-rigid set of rules. When your data are 3 sigma, you get to say "we saw evidence for thus and such." In order to say "we observed something," you must have 5 sigma evidence—this is an extremely high standard.

People sometimes misuse these words. Note that a claim of 3 sigma evidence doesn't mean that there are 739 chances out of 740 that you found something. After all, that's awfully close to 1. I'd bet a fair bit of money on a 739/740 gamble. What it means is that if you did the experiment 740 times, you'd expect to see the kind of evidence you did once on average, even though nothing new was happening.

So, with the example of rolling dice, if we rolled twelve dice, and they came up with ten sixes, this is extremely rare (but not impossible). However, if we saw this, we'd be justified in claiming that something was probably not right with those dice.

The Drama Unfolds

Now we get back to physics and drama. In late June of 2012, both CMS and ATLAS unblinded themselves to their data. Researchers looked at many possible decay patterns of the Higgs, decays into tau leptons, bottom quarks, W bosons, Z bosons, and photons. We recall that the decay of Higgs to photons is rather rare—about 0.2% in the mass region that eventually proved interesting—but the fact that the signal was very striking made it an attractive way to search. In much the same way, the decay to Z bosons is also very attractive in the search for the Higgs. While the decay of the Higgs bosons into Z bosons was more common than decay to photons, the Z bosons themselves are unstable and thus we never see Z bosons directly. Instead, we look for their decay products. One signature that is easy to identify is when Z bosons decay into leptons. So, the analyzers restricted themselves to cases in which the Z bosons decayed into electrons or muons. Under this self-imposed restriction, the decays of Higgs bosons into Z bosons with the desired decays were quite rare at only 0.1%.

It was never expected that the ICHEP conference in Australia would be as exciting as it was, but the LHC experiments were starting to see statistical sigmas in the high fours (recall that 5 is the threshold for "observation of"), which meant that with a little push here or a clever idea there it might be possible to cross that magic number and claim discovery at the conference.

It is true that science is ultimately an impartial endeavor. We use our best techniques to get at the truth and we're all happy to know a little more about the secrets of the universe. But science is also a human endeavor, no matter how much we try to paint it as a bunch of Spocks, arching our collective eyebrows and saying "fascinating." High-quality scientists are competitive. Don't let anyone tell you any different. Neither CMS nor ATLAS wanted to have to clap politely while the other experiment claimed discovery.

Now, this doesn't mean that evidence would be manufactured or claims stretched. High-quality scientists might be competitive, but they're also high quality. They're disciplined. They're careful. Still, they got a lot less sleep that last week.

ATLAS and CMS are big experiments, each with some 3,000 physicists. The experimenters are scientists, sure, but, as I said above, they are people too. They talk with their families and non-collaborating colleagues. It was inevitable that the word of an impending discovery would get out. Then the bloggers woke up.

With all due apologies to the reader, scientific bloggers, in my not-very-humble opinion, can be a real pain. While they are often excellent scientists, they are not involved in the analysis and therefore able to speculate freely. They pass along rumors and exaggerations. Excitement is a substitute for caution. And they are sometimes irritatingly well-informed.

The bloggers are often the tripwire for the more traditional press. The bloggers announce well-informed, but incautious, rumors. The press reads the blogs and gets wind of something and start nosing around, asking the people running the experiments for confirmation, leading them to sometimes say something before they might have done otherwise. I have certainly seen some reject the hype and stand firm. This is what happened in the Higgs case. But there is no denying that in this social-media-connected world there was tremendous outside pressure and scrutiny that scientists in a gentler era would not have encountered.

With ICHEP coming up, the CERN experiments needed to decide what they could say. Did they have a 5 sigma discovery or not? If the data were 5 sigma, then there was another complication. On the one hand, discoveries of that magnitude are almost always announced at the home laboratory first. That meant a press conference at CERN. On the other hand, if there was an announcement at CERN, it would have really taken the wind out of the ICHEP conference, the organizers of which were scientists of considerable stature.

While both experiments kept their results from each other, they both kept the CERN director informed. After all, he is ultimately responsible for the scientific output of the CERN laboratory. He knew that both experiments were very, very close to 5 sigma. So it was decided that what would happen is that there would be a worldwide press conference at CERN in the morning. This timing allowed it to coincide with the opening ceremonies at the ICHEP conference in Melbourne. For Americans, this was dreadfully inconvenient, as it happened in the early, early morning hours of July 4 U.S. time. But it couldn't be helped.

At Fermilab, we opened the laboratory to the public at 2 a.m. and managed to attract several hundred science enthusiasts, some of them still in their pajamas.

The seminar was given in the CERN main auditorium and getting into it was nearly impossible. Some students slept outside the auditorium to get the few seats that weren't reserved for dignitaries. The audience included four of the six people who are most usually given credit for inventing the Higgs mechanism: François Englert, Peter Higgs, Gerald Guralnik, and Carl Hagen. Tom Kibble couldn't attend and Robert Brout had tragically died a year prior and couldn't see his predictions validated. Just prior to the announcement, Peter Higgs was traveling in southern Europe and might well have skipped it. However, a phone call from the head of the CERN theory group changed his mind. John Ellis's phone message was quite clear, "Tell Peter that if he doesn't come to CERN on Wednesday, he will very probably regret it."

In the seminar, the CMS spokesman Joe Incandela presented his experiment's results for an hour, followed by ATLAS spokeswoman Fabiola Gianotti. She good-naturedly complained about going second, because all of the good jokes had already been taken. Both speakers did an excellent job in cautiously announcing what they had discovered and what they hadn't and said the following. They had 5 sigma evidence for a new boson that was *consistent with being the Higgs boson*. That careful phrase means that a boson not predicted by a Higgs-less standard model was discovered and further all evidence was consistent with its being the Higgs boson but we couldn't definitively rule out all other options. The CERN director was a bit more exuberant, exclaiming, "I think we've got it." An example of the kind of event displays—pictures of collisions—that were used in the discovery announcement can be seen in figure 6.11.

It's perhaps worthwhile to say just a bit about what was announced. CMS and ATLAS told a very similar story, both in their 2011 and 2012 data. So there were four consistent measurements. Each of those four measurements (except one, ATLAS 2012) looked at five different ways in which the Higgs boson could decay and several ways in which it could be produced. Every measurement told a consistent story, when one took into account the fact that real measurements have some statistical fluctuations. There was no question that a Higgs-like boson was found.

In addition to the Higgs-like boson, other observations could be pointing to some new and unexpected physics. For instance, both CMS and ATLAS observed more decays into photons than are predicted by Higgs theory. The uncertainties were big enough that nobody got too excited (well, except for some theorists who are always looking for chinks in the standard model armor). And the first measurement by CMS of Higgs bosons decaying to tau leptons seemed to see no evidence for that particular decay chain. The measurement uncertainties

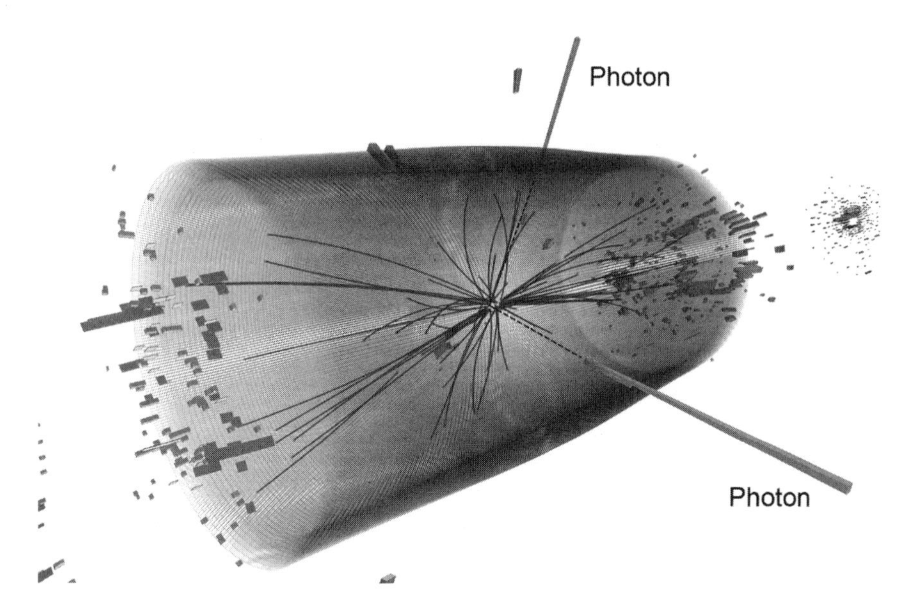

Figure 6.11. This is a display of an event in which a Higgs boson may have been produced and decayed into two photons. Figure courtesy of CERN and the CMS collaboration.

were very large, so that wasn't a big deal either. Still, it generated a little excitement. (After all, if the measurement in the tau lepton decay channel had the same value, but with much smaller uncertainties, it would have killed the Higgs theory quite definitively.)

The press coverage of the Higgs boson announcement was huge. It has been great to see how the people around the world have taken an interest in the microrealm. But, it's not so incredibly surprising, either. People have wondered about the big questions for as long as people have been writing and no doubt long before that. I have given hundreds of public lectures since then, usually to packed houses. I mean, it's not filling a football stadium or anything, but it's not at all unusual to bring together over a hundred people to hear about what could be considered a fairly esoteric topic. I really hope to see that continued interest as the LHC explores even deeper into the origins of the universe.

All the Data

The announcement of July 2012 was that "a new particle was found that was consistent with a Higgs boson." The news reports sometimes retained that caution, although some simply claimed "Higgs boson found." However knowing that what was found was the Higgs boson is actually pretty tricky. Higgs theory predicts the charge, spin, and parity of the particle, although it couldn't predict its mass. However, once the mass was found, the theory predicted all other properties, for instance production and decay rates.

As we have seen, the original measurements of the 2012 data from the LHC studied production and decay modes. Decays of the Higgs boson into pairs of photons, Z bosons, W bosons, tau leptons, and bottom quarks were announced in July 2012, with the photons and Z bosons being the most precise measurements. However, recall that this was with only a fraction of the data recorded in 2012. Surely, the measurements are now more precise and the picture is much clearer?

While there were several updates throughout the fall of 2012, in March of 2013 both CMS and ATLAS announced the results of new studies, this time using all of the data. These final data sets were two and a half times larger than the original ones, resulting in more complete studies of the production and decays of the newly discovered particle. These new studies told basically the same story as the original data, although with reduced uncertainties. However, there was nothing fundamentally new from these later measurements.

What had been missing in July 2012 were measurements of the spin and parity of the new particle. The Higgs boson was predicted to have a spin of zero and positive parity. "Spin" kind of means what it sounds like, as it has a connection to particles spinning or moving in an orbit. This is what physicists call *angular momentum*. However, in the quantum realm, spin differs from the familiar, macrolevel (i.e., human sized) concept of spin. For instance, in the world of the big, particles can spin fast or slow or anywhere in between. In contrast, in the tiny world of subatomic particle physics, spin can only take on values in multiples of one-half: 0, ½, 1, 3⁄2, 2, and so on. Values of ¼ or 2.1 are forbidden.

From the various decay modes observed in 2012, we knew that the spin of the new particle was either 0 or 2. If it turned out to be 0, this would be support for the Higgs hypothesis. If it turned out to be 2, we didn't find the Higgs boson.

Parity has to do with what happens when you swap left for right, up for down, and forward for backward. If you make those swaps and can't tell the difference, the parity is said to be positive. If, when you make the swaps, it looks the opposite, the parity is said to be negative. This is a little tricky to visualize, but you can get the most important points by looking in your bathroom mirror. Mirrors swap left for right. So suppose you stand in front of the mirror with your left arm raised and your right arm down. The mirror image is the opposite, with the right arm raised and the left arm down. This is an instance of negative parity, as you and the image are opposites. However, you could equally well raise both your right and left hands. Your reflection also has right and left hands raised. Since you and the image are the same, this is a case of positive parity. Figure 6.12 illustrates these ideas.

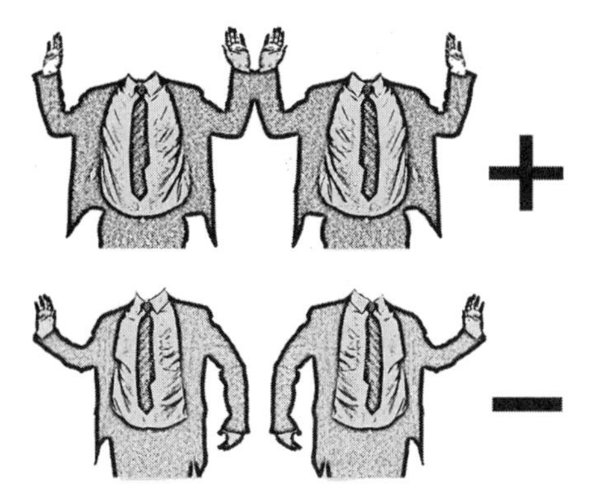

Figure 6.12. The concept of parity can be illustrated by looking at a person in a mirror. A reflected image has a reversed sense of left and right. If a person raises both hands (*top*), the reflected image also has both hands raised. Thus the reflection causes no change, and we call this *positive parity*. In contrast, a person with a raised right hand and lowered left hand (*bottom*) will have the opposite hands raised in a mirror. This is called *negative parity*.

Both ATLAS and CMS studied the spin and parity of the new particle. The way you determine the spin of a particle is by looking at the angles at which the decay products come out. You can understand this by thinking of a particle with no spin in empty space. Since space is empty and there is no axis around which the particle is spinning, all directions are equivalent. There is nothing to make one particular direction special. Therefore, particles with no spin must decay in all directions with equal probability. However, a spinning particle has an axis around which it spins. This axis is now different from all other directions. Because of this, certain decay directions are preferred over others. This idea is illustrated in figure 6.13.

Using sophisticated techniques, both experiments studied the spin of the new particle and, with very high probability, it was determined to be zero. Using similar techniques, it was also shown that the new particle had positive parity. Both the spin and parity were found to be the same as that predicted by Higgs theory.

So, we now know that the particle found at the LHC is electrically neutral, with zero spin, positive parity, and production and decay modes consistent with the theorized Higgs boson. Is that sufficient to declare that the Higgs boson was found? Rather surprisingly, the answer is actually no.

As I mentioned earlier, Higgs theory predicts that there were four Higgs particles but that three interacted in such a way to make the physical W^+, W^-, and Z bosons, leaving one and only one undiscovered Higgs boson. However, there are other theories that predict more. One such theory (supersymmetry,

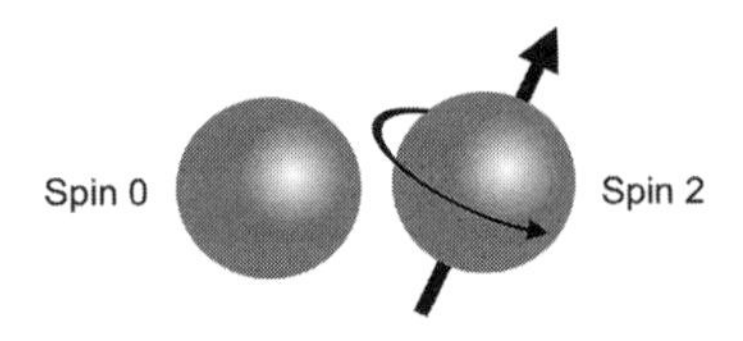

Figure 6.13. If a particle has no subatomic spin (*left*), there is no preferred direction. This means decays must occur in all directions with equal probability. In contrast, if a particle has some spin, the axis of spin defines a "special" orientation, which can lead to preferred decay directions (*right*).

discussed in the next chapter) predicts not one but at least five physical and observable Higgs bosons. And, to make things tricky, one of those five bosons has the same properties as the Higgs boson of the standard model. Thus, even if we knew that the new particle had exactly the same properties as the Higgs boson, it is possible that what we found is simply one of the five Higgs bosons of supersymmetry and just hadn't found the other four. It is for this reason that in the summer of 2013 we physicists said that we found *a* Higgs boson but are unwilling to say that we found *the* Higgs boson. To say *the* Higgs boson, we'll have to prove that the reason that we didn't find the other four is that they're not there. That's very hard to do and it is impossible without a lot more data. Thus, until we begin collecting data again, the situation remains murky.

So where do we stand now? As of this writing, the data continue to support the premise that the LHC experiments have found a Higgs boson, perhaps the Higgs boson. It is certainly looking like those papers in 1964 by Brout, Englert, Guralnik, Hagen, Higgs, and Kibble were on to something. Even if there is more to the Higgs mechanism than those simple ideas of fifty years ago, there is sufficient agreement between the predictions and the data for a consensus to have formed that Higgs theory is an important piece of the puzzle.

Given the impact of the theory and the magnitude of the discovery, it was clear that it was time to start thinking about a phone call in October 2013. I'm talking about, of course, the annual ritual whereby one or more phone calls go out notifying a lucky few scientists that they have been awarded that year's Nobel Prize.

But who should get the Nobel for the Higgs boson? There were (at least) six physicists who could lay a credible claim to the original theoretical idea. On the experimental side, there were a lot of contenders, from the "corporate" entities like the CERN laboratory itself to the ATLAS and CMS experiments. There were many people who contributed to the discovery, from the experiment leaders to the young graduate students and postdocs who did the yeoman work of analyzing the data. If the committee awarding the Nobel Prize decided to pick the leaders, there is the difficulty that the accelerator and detector took something like two decades to design and build. Should the people with the original ideas be recognized? The ones who built the equipment? The ones who happened to be in charge when the discovery was announced? Who?

The Nobel Prize has rules associated as to how many people can receive them. They have evolved a little over the years, but the current rules are:

> A prize amount may be equally divided between two works, each of which is considered to merit a prize. If a work that is being rewarded has been produced by two or three persons, the prize shall be awarded to them jointly. In no case may a prize amount be divided between more than three persons.
>
> Work produced by a person since deceased shall not be considered for an award. If, however, a prizewinner dies before he has received the prize, then the prize may be presented.
>
> Each prize-awarding body shall be competent to decide whether the prize it is entitled to award may be conferred upon an institution or association.

According to these rules, the recipient must be alive and there can be no more than three. The committee can decide whether to make the award to people or institutions.

As the 2013 Nobel season rolled around, the buzz began to award the Nobel Prize for the Higgs boson. Because of the "rule of three," it was clear that not all the deserving theorists could be awarded. Brout and Englert published first, Higgs second, and Hagen, Kibble, and Guralnik last. By rule of precedent, Brout and Englert could be sensibly awarded the prize, but it would have been awkward for Peter Higgs not to get a prize for the prediction of the Higgs boson. And Brout died in 2011, becoming ineligible for the honor.

I attended the 2013 European Physical Society conference in Stockholm and even had dinner with Higgs and Englert. The rumor mill buzzed, with an unofficial consensus starting to form that Peter Higgs and Francois Englert were the frontrunners. The other three theorists would break the rule of three, so that seemed to be that. The only question was whether there might be some other person suitable for the third slot. I had thought that perhaps Lyndon Evans, lead scientist to build the LHC, might be a contender, but that idea never seemed to be a popular one. Another idea that was floated was the third slot to go to CERN itself. While the physics Nobel Prize has never gone to an organization, it is within the rules and there is always a first time.

The Nobel Prize in physics was scheduled to be announced on October 8, 2013, at 11:45 a.m., Stockholm time. I was in Chicago and dutifully connected to the Web cast at 4:45 a.m. to see who the winner would be, only to find that the announcement was delayed and then delayed again. What could be wrong?

At 6:15 a.m. Chicago time, the committee made their announcement. As was generally expected, Englert and Higgs won the Nobel. Englert had been contacted, but Higgs was nowhere to be found. He had told everyone that he was going on vacation that time because, win or lose, he didn't want to deal with the media onslaught. When you're a crotchety, 84-year-old genius, you can do that sort of thing.

The origins of the delay are not completely clear. Some said that it was simply due to the committee's inability to contact Higgs. Others claim that the discussion went on for some time to decide on whether it was time to recognize organizations and use the third award slot for CERN. I wasn't in the room and cannot personally confirm or deny any rumors, although individuals with better connections than I have told me that both reasons played a role in the delayed announcement.

The decision to recognize the prediction of the Higgs field and boson was an apt one in my opinion. The theory was strong and confirmed enough for the award to be fully credible. The theorists were approaching very advanced age and waiting another decade would have been ill-advised. The Nobel Prize is the pinnacle of scientific achievement, and the committee made the right choice. I do feel sorry for Hagen, Kibble, and Guralnik, as they are smart men and they made important contributions as well. But in science there are few awards for third place. And life moves on.

We should leave this topic with one parting thought. You'll recall that the Higgs field was added to the electroweak theory as a Band-Aid and didn't originate in some deeper theory. So, it's great and all if the Higgs theory happens to be true. But we have no fundamental understanding of why. For that, we'll have to dig deeper into the mysteries of the universe. Perhaps some young reader of this book will be the one who cracks that particular nut. In the next chapter, one of the things we'll talk about is what direction a Higgs boson with a mass of 125 GeV takes us.

7

LOOKING FOR SOMETHING NEW

Before we proceed into this chapter, I should warn you that everything included here is completely speculative. We've left the comforting confines of what we know and have leapt into the unknown. At the edge of human knowledge, there is never certainty. Indeed what we find in our experiments at the LHC may be similar to what we discuss below, or it may be something entirely different. Keep this in mind as you read. But this chapter does give you a good idea of some of the interesting questions and some of the things we think that we might find in the years ahead.

While we know a lot about our universe, no one would argue that we know it all. Let's briefly recap what we do know and see what sorts of questions are raised.

The observed universe is composed of two types of particles: quarks and leptons. Quarks are affected by all of the four forces: strong, electromagnetism, weak, and gravity. Leptons are not affected by the strong force, and a subclass of electrically neutral leptons, the neutrinos, is not affected by the electromagnetic force either. We also know that there appear to be three identical "generations" of quarks and leptons, with each generation containing heavier copies of similar quarks and leptons.

We also know about the forces. The four forces have very different strengths, with gravity being ten thousand, trillion, trillion, trillion (about 10^{40}) times smaller than the strong force. Some forces are attractive, while others both attract and repel. Each of the forces—except gravity, which is theorized to have a particle associated with it, named a graviton, but that particle has not yet been found—has been shown to be caused by the transfer of subatomic particles, called photons, gluons, and the W and Z bosons. These particles can be electrically charged or neutral and have either zero or considerable mass.

Another fascinating piece of the story of forces is historical. In the past, our understanding of the nature of the world was less advanced than it is now. People saw that things fell when you dropped them. They also saw that the sun rose and set, the moon had phases, and the seasons came and went. These things seemed to be unrelated, until a young genius by the name of Isaac Newton showed that the cause of all of these phenomena was gravity. We could say that Newton "unified" the behavior of falling things and the motions of the heavens with a single principle that explained them both.

Similarly, while people have been aware of static charge, lightning, magnetism, and light for millennia, it was only in the 1800s that all of these phenomena were shown to be a single thing, now called electromagnetism. Recall from chapter 2 that in the 1960s physicists were able to show that electromagnetism and the force governing some kinds of radioactive decay (the weak force) were actually the same thing. Particle physicists now speak of the "electroweak" force. The Higgs boson plays a central role in linking electromagnetism and the weak force.

This brief historical interlude leads us to the following question: While we speak of four forces—strong, electromagnetism, weak, and gravity or three if we speak of electroweak—is it possible that further study will reveal that these seemingly unrelated phenomena are really all the same thing?

So, with these thoughts in mind, let's ask some more questions:

- Why do the forces have such disparate strengths and ranges?
- Do the observed forces end up being different ways to observe a single principle? If so, at what energy will their common origins be made apparent and why is that the energy at which the commonalities are evident?
- Why quarks and leptons? Why do some particles have mass and others don't? Ditto electric charge? Why are quarks the only matter particles that feel the strong force? Why are there three generations? Could there be other generations?
- We live in a universe with three spatial and one time dimension. Why? Could there be more of either type of dimension? If so, what would they look like and, if they exist, why haven't we seen them?
- Why is the universe made only of matter, when we make matter and antimatter in equal quantities in our experiments? Where did the antimatter go?

There are other questions on which the LHC is expected to be silent or comment on only indirectly. But the LHC is designed to explore the questions listed above (and many, many more), as well as to accurately measure familiar things at the higher energies that only it will provide (like how quarks scatter, for instance).

A book like this cannot possibly address all possible questions. Thus we will restrict the discussion to a few major topics. They are the following:

- Are there any further mysteries that come along with the discovery of a Higgs boson?
- Will all the forces be shown to actually be the same thing? Why do current experiments hint that the energy at which the forces might unify is so high?
- Why are there generations, and do they signal that there is something smaller inside quarks and leptons?
- Are there more than the three space and one time dimensions that we are familiar with?

Finally, two additional questions will be discussed in this chapter, but they will be given lesser attention. The reduced attention doesn't mean that they are of lower importance (after all, I've skipped some very important questions) but rather indicates that the LHC is not the only facility addressing them. However, because the LHC hosts two special purpose detectors with specific goals, their mission needs to be mentioned here. One detector at the LHC complex is designed to intensively study particles with the hope of shedding light on why we don't commonly see antimatter in the universe. Another detector looks into what happens when lead nuclei are slammed together at high energy. These collisions will provide insight into what happens to matter when it is heated enough to allow quarks to freely escape their proton and neutron cocoons. It will also explore the conditions during a period of the early universe about which we are currently largely ignorant.

Keep in mind that it is completely wrong-minded to say that "the LHC was built to discover X." That would mean that "X" is understood well enough to know that it's there and therefore finding it isn't really a discovery. No, the purpose of the LHC is studying the nature of matter under conditions that are seven times hotter and more energetic than ever before observed at earlier accelerator facilities. We will see what we see: interesting, fascinating, or disappointing; the universe will reveal some of its secrets and the world will become slightly less of an enigma.

However, scientists could not have persuaded the world's funding agencies to support a multibillion-dollar endeavor if they didn't have a very good reason to expect that there *would* be valuable discoveries. Probably the most likely and anticipated discovery for the LHC's experiments is the explanation of why subatomic particles have mass. We talked in some detail in chapter 6 about the Higgs boson and the particle discovered in the summer of 2012. While that was undoubtedly a huge advance in our knowledge, with it came a considerable mystery. This mystery has to do with the amount of mass the Higgs boson has and is generically referred to as *naturalness*.

Naturalness

You may recall one of the ways we determined that we found a particle consistent with a Higgs boson was by finding a particle with a consistent relationship between its mass and the predicted Higgs boson decay rates. You may also recall that people are buying me drinks because I guessed it might weigh about 125 GeV. It turns out there is a mystery associated with a Higgs boson with a mass of 125 GeV. This is a bit subtle, so we'll take it slow. It starts with the question of where the mass of the Higgs boson originates.

The Higgs field is what gives particles their mass, as we saw in the previous chapter; so, since the Higgs boson is massive (in fact, it is the second-heaviest fundamental particle discovered), it must interact with the Higgs field. That's perfectly fine. But the world is more complicated than that. There are considerations that arise from the weird world of quantum mechanics that have to be taken into account. For instance, no subatomic particle sits there without changing its identity. So, the Higgs boson spends part of its time being other particles. For instance, it can temporarily turn into a top quark/antiquark pair or into a pair of W bosons, before they coalesce back into a Higgs boson. In fact, it can turn into pairs of lots of different particles (figure 7.1); however, since the Higgs boson likes heavy particles by interacting more with some particles and making them heavy, only a handful of particles matter. The subatomic particles of chapter 2 that are heavy enough to matter in this context are the W (80 GeV), Z (91 GeV), and Higgs (125 GeV) bosons and the top quark (173 GeV). The next heaviest particle is the relatively piddly bottom quark (4.5 GeV). While the bottom quark comes into the mix, its effect is negligible. Consequently, the standard model predicts that only these four biggies really matter in the process of the Higgs boson temporarily transforming itself into other particles.

What is the mystery? To get an inkling as to what it is, you need to realize that, when theorists calculated the mass of the Higgs boson that was expected to be observed, they had to first calculate the mass of the Higgs boson as if the modifications to the theory that come from quantum mechanics—the effects of other particles outlined above—didn't exist. Then they needed to add in the effect of any quantum fluctuations. It's important that the quantum effects are small. If they were huge, then the predicted mass of the Higgs boson would be huge. I know I said that the Higgs is the second-largest particle, but recall that gradation is in the tiny realm of subatomic particles. When I say "huge" here, I mean really huge, like a hundred quadrillion times heavier than we observed. Since the boson discovered in July 2012 is (1) probably the Higgs boson and (2) has a pretty small mass, the quantum effects on the boson need to be somehow kept in check.

The prediction of the size of the quantum effects has two parts. The first part

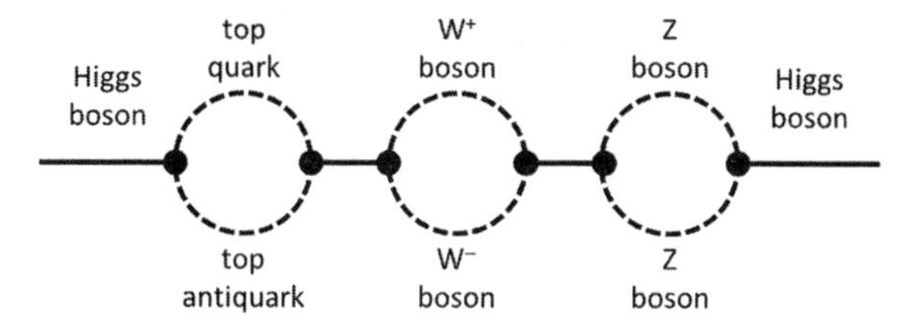

Figure 7.1. Higgs bosons can spontaneously convert into pairs of other subatomic particles and then back to a Higgs boson. These pairs exist only for a very short time, but their existence will alter the observed mass of the Higgs boson.

is the highest energy that the standard model is expected to apply. (That's just a way of saying the energy at which the standard model breaks down and must be replaced by a more comprehensive theory.) In the standard model, that energy is huge. In order to balance out that huge energy, we need to have the second effect to be very small to cancel things out.

This second effect involves the masses of the particles that cause these changes in the mass of the Higgs boson due to quantum mechanics. These are the four biggies mentioned above. We need them to combine so that the combined effects of quantum mechanics is nearly zero. That will get rid of the huge energy term. The way that these particles combine is that the bosons (Z, W, Higgs) contribute positive numbers, while the fermions (in this case a top quark) provide negative contributions to the changes caused by quantum mechanics. Since you can add positive and negative numbers and get zero, that's good. (If you are getting a little lost here, bear with me. All of this will come together soon, I promise.)

However, we don't have any good reason to for these numbers to "just happen" to add to zero. In physics, things never "just happen." There has to be a reason. Since we don't know the reason, that means that there is some new theory (still to be discovered) that will tell us why the cancellation is "natural." Because what we see now is just not natural.

To see why we call this conundrum the "naturalness problem," let's use an analogy of a highly successful grape farmer and the grocer. The grape farmer knows that he is going to have an amazing year of grape production. So he tells the grocer that he needs to build a huge storage facility but doesn't tell him more than that. The grocer then builds a building that he thinks will hold all the grapes. The building is a cube a hundred feet on a side. Such a building can hold exactly 33,428,546,008,198 grapes. When the grape farmer shows up to deliver the grapes, the most likely thing is that the building is the wrong size. After all, most grape producers might deliver five or ten truckloads of grapes and consider that an amazing year. In that case, the building would simply be too large. Perhaps the grocer didn't understand that the grape grower was actually a representative of Grape Growers International and what he was talking about was delivering the world's grape production. In that case, the building is likely too small and there would be grapes heaped up in piles all around the property.

Now, suppose, seemingly against all odds, without any more information than "an amazing year of grape production," the grocer's building was exactly, precisely, the right size. The grapes, when delivered and stored, fit into the building with no spare space and no grapes left outside. That would just be an extremely weird coincidence. The only way you could get such a perfect agreement is if the grocer were given utterly, ridiculously, exact instructions. This is kind of like the balancing necessary in the Higgs boson case, except to make the comparison fair the warehouse would have to be five million miles on a side. There's just no way for that kind of grape production and storage to be so exactly matched.

In a similar way, the cancellation of the huge energy term in the Higgs quantum corrections needs to be nearly perfect. Perfect cancellation can only occur if some physical principle is enforcing it. This means that the low mass of the Higgs boson essentially ensures that a discovery is out there. This discovery will explain how the cancellation that seems so unnatural to us is actually exactly what we should have expected.

Introducing the Fudge Factor

Let's talk about this problem in a little more detail. In order to be able to calculate the observed mass of the Higgs boson, you have to take into account the fraction of the time that it exists in other forms. This alters the mass of the Higgs boson just a little bit. It's easy enough to calculate the effect these ephemeral particles have on the mass of the Higgs boson. (OK, OK . . . it's actually rather difficult to calculate, but it's a piece of cake for my theoretical colleagues.*)

The brave can peer at the footnote, but we can simplify the bigger picture the following way. Not sweating some squares and ignoring some important constants in the equation (which one can't really do, but it's OK for big picture thinking), we can say that the

> (observed Higgs boson mass) = (theoretical Higgs boson mass) + (fudge factor).

* The actual equation is the following: mass(Higgs observed)^2 = mass(Higgs theoretical)^2 + $[k\ \Lambda]^2 \times$ [mass(Z boson)^2 + 2 × mass(W boson)^2 + mass(Higgs observed)^2 – 4 × mass(top quark)^2]. In this equation, k is a technical constant and Λ (a capital Greek lambda) is the maximum energy that the theory applies.

When these quantum fluctuations aren't considered, the fudge factor is zero. Of course the quantum fluctuations do exist, and the fudge factor has these basic properties:

(fudge factor) = (highest energy at which the theory applies) × [(effect due to bosons) – (effect due to fermions)].

The essence of the naturalness problem stems from the fudge factor. You can see from the equation above that it is composed of the maximum energy for which the theory applies times a number made up of some positive numbers added to some negative numbers. Let's explore that a bit. To do that, let's write down what we know in an abbreviated way. The theory says:

[observed] = [theory] + [maximum energy] × [mass stuff].

For the purposes of this discussion, let's assume that the boson we found in July was the standard model Higgs boson. This means that the Higgs boson has a mass of about 125 GeV. If the product of the maximum energy and mass stuff is a number like zero or five or ten or even 100, this isn't a problem. That means that we can credibly add the theoretical mass and the extra term and get something near 125 GeV. However, if one of the energy or mass stuff terms is large, then we may have a problem.

In the standard model, there is no new physical phenomenon between the stuff we're studying at the LHC and 10^{19} GeV, which is called the *Planck scale*. This is the energy at which the theory says that gravity becomes strong and at which point it is expected that new physical laws begin to apply. This is in principle the highest energy for which the standard model is valid. So the maximum energy term is huge. If you scan back a few sentences, or bravely read the footnote, you'll note that I said I was ignoring the places where the equation squared some numbers. One of those ignored squares was in the maximum energy element of the equation. The maximum energy term is 10^{38} (which is a 1 followed by 38 zeros). That number is utterly huge, and we can get a good feel for that by putting it into our abbreviated equation:

[observed] = [theory] +
[100,000,000,000,000,000,000,000,000,000,000,000,000] × [mass stuff]

This means that the [mass stuff] term has to be very nearly zero. It can't be 0.00001 or 0.0000000001. Neither of those tiny terms can tame that humungous number. The effect due to the masses has to be incredibly, incredibly, small.

Let's remind ourselves what goes into the [mass stuff] term. It is the contribution to the mass of the Higgs boson from quantum mechanics caused by bosons minus the contribution caused by fermions. When we explicitly state the particles with a mass big enough to matter (and again ignore some places where terms in the equation are squared, plus a stray number here and there), we can write the [mass stuff] term as

[mass stuff] = [(Z boson) + (W boson) + (Higgs boson)] – (top quark).

So that kind of makes the whole thing pretty clear. We know that [mass stuff] has to be nearly zero. All the positive terms are bosons and there is a negative fermion (top quark) term. It's possible to add positive and negative numbers and get zero, so we're good. Except. . .

Except that we have no understanding of why particular particles have the particular mass that they do. We know that they get mass through interacting with the Higgs boson, but just why it is that the bottom quark has a smaller mass than the top quark, and what determines the exact numerical value of the mass of each particle isn't known.

We now come to the origin of the term "natural." It is utterly unnatural for the effect of the bosons and the effect of the fermions to "just happen" to be so perfectly balanced, as with our theoretical grape harvest fitting exactly into the storage building. If you have a calculator handy, you can use the equation from the footnote and the values for the relevant particles, and you'll see that the necessary cancellation just doesn't work: mass(top) = 173.07 GeV, mass(W) = 80.39 GeV, mass(Z) = 91.12 GeV, mass(Higgs) = 125.0 GeV.

Ideally, a physics theory shouldn't have any places where you look at a phenomenon or cancellation and say "wow . . . that's a weird coincidence." Instead, one should be able to look at the theory and say "well, *of course* those things cancel; it's because these two ideas are interconnected." However, we don't have that here. We just see that the [mass stuff] is small for some unknown way. It's just completely unnatural and hence we call it the "naturalness problem."

There is no known answer to the naturalness problem. But many physical theories are being considered that can provide answers to this thorny issue. In the next pages, we will talk about supersymmetry, extra dimensions, and the possibility that quarks and leptons are composed of smaller particles. All of these theories are interesting in their own right. They each can potentially (recall that none have been established yet) answer other questions listed at the beginning of the chapter. However, each of them also has a potential answer to the naturalness problem. Thus as we close our discussion for each theory, we'll spend a little time explaining its particular answer to this as-yet-unsolved mystery.

It is worth mentioning that we've focused more on the cancellation of the [mass stuff]. However, it is possible that we were wrong about the maximum energy at which the standard model applies. Maybe the standard model breaks down not at the Planck energy, but at energies just ten or a hundred times more than the LHC can access. If that's the case, that's great too; the reason is that, if the standard model is breaking down, this means some other theory needs to take over.

No matter what—whether the prediction of the maximum energy at which the standard model applies is wrong or whether we will discover something new that predicts the mass cancellation—we need some new physical phenomenon to tame the mass of the Higgs boson.

Bottom line? If we've discovered the Higgs boson and its mass is 125 GeV, this is completely unnatural in the standard model and we *must* discover some new physics to explain it.

Supersymmetry

There are mathematical principles that govern the universe and that are embodied in the equations that physicists use to calculate their predictions. The most critical of these principles are called *symmetries*. There are many kinds of symmetries, some well-established and with which the reader is very familiar and some rather counterintuitive and about which the LHC is hoped to say something profound. Before we can understand this particular LHC goal, we need to spend some time defining some basic concepts.

The first idea is that of symmetry itself, a concept that we touched upon in the Higgs boson section in the previous chapter. Symmetry is both a mathematical as well as a visual or artistic concept. Basically in both math and art, a symmetry is something you can change and nobody will know.

A circle is the most symmetric two-dimensional object. (We pick two-dimensional because this page is two-dimensional. We could as easily use three-dimensional, but then the symmetric shape would be a sphere.) In figure 7.2, we see that when we start with a circle, you can flip, rotate, move, or look at it again tomorrow and you can't tell the difference. Technically, we say that the circle is "symmetric" under all these possible changes. For the more mathematically inclined, we'd say that the shortest distance between the center and a point on the perimeter is unchanged under these operations.

In physics, symmetry has a similar meaning but an added physical significance. Imagine a bug at the bottom of a bowl. In order to crawl out, he'll have to do a lot of work. If he crawls out by going to the right, he'll have to do some amount of work. If he tries to crawl out to the left, it's the same thing. In fact, and here's the important thing, if you just changed the words "right" and "left" in the last two sentences, nothing would change and all conclusions you would draw would be unchanged. Similarly, if you moved the bowl across the room or moved it from the table to the floor, or vice versa, the bug's predicament would

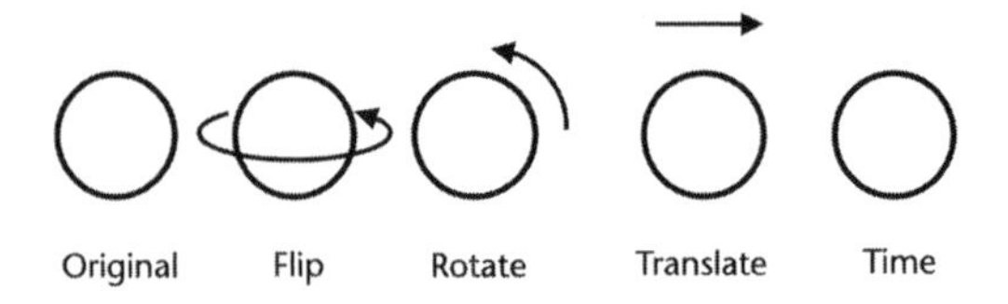

Figure 7.2. A circle is the most symmetrical two-dimensional object. Do nearly anything to them and you can't tell that something happened.

be identical. And, assuming that he brought a snack and a little bug sleeping bag, so he'd be well fed and rested, his effort needed to get out would be the same tomorrow as well.

Getting beyond examples of bugs and bowls, physicists can write equations describing the standard model, which you can recall is the theory that embodies our current understanding of the universe. In fact, the standard model equation includes all possible symmetries save one. This additional symmetry deals with how particles spin at the quantum level. Before we finish introducing this interesting new symmetry, we need to learn a couple of things about spin and quantum mechanics.

We are accustomed to thinking about certain aspects of the world as being "quantized," or coming in discrete units. All electrons have a single, identical, mass. It is impossible to have an electron with half the mass of the others. Similarly, electrons only have one charge. Electrons do have another feature that is not as quite as familiar. Every electron in the universe is spinning identically. This is counterintuitive, since we are used to objects being able to spin faster or slower and not having a single amount of spin that is allowed. However, we're also accustomed to being able to change mass at will. A wheelbarrow of sand can have more or less weight, depending on whether we toss in that final shovelful. And yet each electron has an identical mass. So the disparity between the ideas of spin at the classical level and the quantum level maybe isn't that hard to accept.

The numerical amount of spin isn't really important. (For the technically minded, the spin of the electron is ½ ħ, where ħ is a constant of the universe and is a tiny number.) We can ignore its numeric value and just recall that all spins are expressed in units of ħ and therefore simply drop the "ħ", calling the spin of the electron ½. Not only is it true that all electrons have spin ½, the same is true of quarks. In contrast, the force-carrying particles, the W, Z, gluon, and photon all have a spin of 1, or twice that of the quark and electron. The Higgs boson has a spin of 0.

Further study has revealed that there are two fundamental classes of particles: the fermions, with "half integer spin" (e.g., ½, 3⁄2, 5⁄2, . . .) and bosons, with "integer spin" (i.e., 0, 1, 2, . . .). You'd think that a little thing like a half unit of spin wouldn't make all that much difference, but it does. Fermions are the rugged individualists of the particle world. No two fermions can be in the same space at the same time and with the same energy. This fact has huge consequences for chemistry, which, after all, is the study of the fermion electrons around atoms. For those who have taken chemistry, this is the source of the Pauli exclusion principle and explains a lot of the structure of the chemical periodic table.

Bosons, in contrast, are gregarious. "The more the merrier" is their motto. Two bosons can be in the same place at the same time and with the same energy. No problem.

Getting back to symmetries, there remains a possible symmetry not yet observed. It is called *supersymmetry* and is the idea that you could exchange fermions and bosons everywhere in the equations (and in the universe) and nobody would notice the difference. Physicists often use the term SUSY as an abbreviation for supersymmetry.

Well, this symmetry has not been officially added to quantum mechanics for the simple reason that it's simply absurd. If we replaced the observed fermionic electrons with a boson equivalent, all of chemistry would be radically different.

Nonetheless, mathematically at least, one can think of constructing a physics theory in which supersymmetry exists and in which you can swap all fermions for bosons, and vice versa. It is important to note that supersymmetry is *not* a theory. It is a *principle*. It's like "conservation of energy" for the scientifically minded or "greatest good for the greatest number" for the philosophers. There are many ways one might try to apply this guiding philosophical principle, with some adopting a "Mother Theresa" behavior to help the poor, while another might adopt a "Bill Gates" behavior and set up a philanthropic foundation. The principle is "greatest good," while a theory would be the Mother Theresa or the Bill Gates individual implementation of the principle.

Similarly, in physics, many specific theoretical models *incorporate* the principle of supersymmetry. But these theoretical models are not supersymmetry, per se. Supersymmetry is much bigger than that.

In 1981, Savas Dimopoulos and Howard Georgi took the conventional standard model and added supersymmetric principles to it. This new model is called the *minimal supersymmetric standard model*, or MSSM. MSSM, as its name suggests, is the usual standard model, with the absolute minimum number of changes necessary to incorporate supersymmetry. It would be easy to overcomplicate what was done, but in essence they added terms to the standard model equation. The standard model has terms for the matter fermions (i.e., the quarks and leptons) and for the force-carrying bosons (i.e., gluons, photons,

Ws, and Zs). The MSSM had these terms, plus two more, that are equivalent to the quarks and leptons except as bosons and the other equivalent to the force carriers, except as fermions.

With the addition of these terms, a most unsettling thing occurred. Just like the first two terms in the standard model meant that quarks, leptons, and the force-carrying bosons exist, the second two terms in the MSSM predicts that more particles exist. In fact, the number of particles predicted by the MSSM is precisely double what we currently know about. The fact that we've never seen these extra particles has led some skeptical physicists to note (with not a little sarcasm) that at least we've discovered half of the particles predicted by the MSSM theory. The MSSM also requires that we have not one, but five different Higgs bosons, the one you're now familiar with and four others. Two of these newly predicted Higgs bosons have electric charge. (Note that these charged Higgs bosons are unrelated to the ones mentioned in chapter 6. The bosons of chapter 6 merged with the bosons involved in electroweak unification and became the W^+ and W^- bosons and the heretofore elusive Higgs.) The bosons described here would be particles that are yet to be discovered and don't fit into the framework described in earlier chapters. Thus if the LHC experiments find more than one Higgs boson, this will be evidence that the idea of supersymmetry has some merit. This is one of the goals of scientists looking for the Higgs boson—determining whether the particle observed by ATLAS and CMS and described in chapter 6 is the single Higgs boson of the standard model or one of the Higgs bosons predicted by supersymmetry.

There are simple naming rules to these newly predicted particles. The bosonic equivalent to our familiar matter particles has the same name, except with an "s" before it. Thus quark becomes squark; lepton becomes slepton, and so on. The fermionic equivalent of the familiar force-carrying particle gets an "ino" added to the end, with occasionally a little phonetic surgery to make the word easier to say. The W becomes a wino (pronounced weeno), photon becomes photino, and so on. In all cases, we can denote a supersymmetric particle by putting a tilde over the symbol, thus the equivalent of a photon is a photino ($\tilde{\gamma}$) and a top quark is a stop squark shown by $\tilde{t}$. Figure 7.3 gives the entire list.

In a moment, we'll get to how we might find supersymmetry in our detectors. But in the meantime, let's discuss why you might want to add terms to your equations that would double the number of particles you predict, with zero physical evidence that they actually exist. As we mentioned at the beginning, there are mysteries in the universe. One of the interesting questions is that of force unification. Just like Newton showed that the thing that keeps my cat firmly placed on my keyboard as I write is the same thing that governs the planets, current physicists hope to show that the four forces of which we are

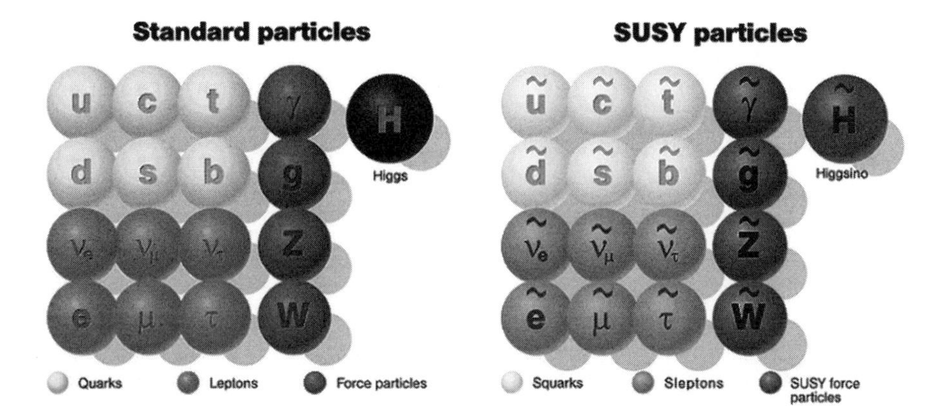

Figure 7.3. The supersymmetric, or SUSY, particles have an identical organization to the normal particles. They are denoted by adding an S to the fermions, an "-ino" to the end of the bosons, and a tilde over their symbols. These particles are part of a principle that says that we could swap out all fermions and bosons and no one would notice; their existence has yet to be proven. Figure courtesy of DESY, Deutsches Elektronen-Synchrotron.

aware are really one and the same. It is not true that this idea is taken as an article of faith, nor do all physicists believe that it is inevitable. But it sure would be elegant if it were true.

In the 1960s, physicists were able to show that the weak and electromagnetic forces were one and the same (and should properly be called the "electroweak force"). The question is "Can we show that the strong force and the electroweak force are just different ways of looking at the same thing?" And what about gravity? The answer to these questions is currently "no," but there are reasons to think that the question is a reasonable one and that the answer may eventually become "yes."

For instance, we can measure just how strong the three quantum forces (strong, electromagnetism, and weak) are. If we measure the strength at different energies (that is to say in collisions of different violence), we see that their strengths aren't constant and that they actually change with energy. If we project the trend of the three forces, we see that they all become the same at the rather high energy of 10^{14} to 10^{15} GeV, or a hundred thousand billion times the mass of a proton. We call this the *grand unification theory energy*. Contrast this to the energy at which the symmetry between electromagnetism and the weak force is broken, which is about 1,000 GeV. The fact that the three forces "just happen" to have the same strength at some energy is suggestive (but not proof) that they are all one and the same thing. The fact that they all merge at one particular energy is very interesting.

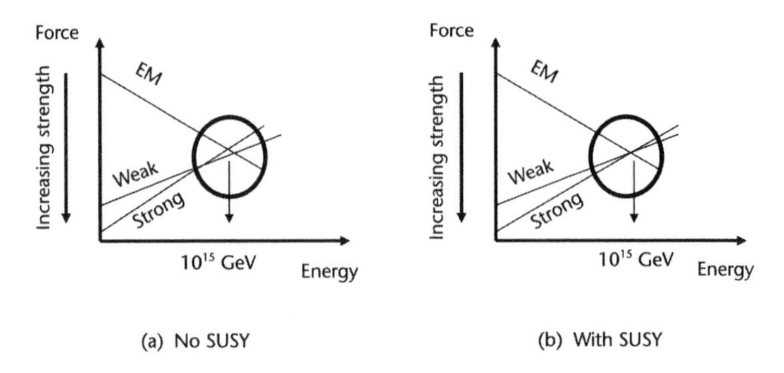

Figure 7.4. A prediction of the standard model is that the strength of the subatomic forces varies with energy. All forces seem to approach the same strength at high energy even under the current theory (*left*), shown by the lines that almost intersect in the circle. It is easy using supersymmetric principles to make the three forces unify at a single energy (*right*). GeV = one billion electron volts; EM = electromagnetic force.

On closer examination, we see that the three forces do not project to *exactly* the same spot in the standard model theory, as shown in figure 7.4. However, and this is a suggestive beauty of the MSSM, it is pretty easy to use supersymmetry to make the three forces merge at *exactly* the same energy. This is not proof that supersymmetry is right, but it gives us a warm and fuzzy feeling nonetheless.

Even without the perfect unification of supersymmetry, there is a nagging question. Why is it that electroweak symmetry breaking occurs at so much lower an energy than the grand unification energy? That's just weird and unnatural. It's kind of like the finances of a billionaire. Every month, she has earnings and expenditures. Large amounts of money slosh into and out of her bank account. If, at the end of every month, her account had under a dollar in it, that would be weird. It's hard to imagine these large million-dollar deposits and withdrawals could balance that perfectly without some principle making it so.

If it turned out that this was her "charity account" and it was set up so that the deposits were carefully designed to cover planned donations that were automatically transferred to the charity, then the bank balance would make sense. But without the "charity principle," it would remain very mysterious that the account would be balanced so well.

Similarly in physics, it's rather odd to have the grand unification scale be a hundred billion times greater than the electroweak symmetry breaking scale. By all rights, they should both be more similar (and nearer the high end). Thus the Higgs boson mass (which, as we learned in the previous chapter, plays a critical role in electroweak symmetry breaking) should be much, much higher than suggested by the data we've taken to date. Perhaps there is a principle that explains this disparity in the unification energies and supersymmetry fits the bill quite well. If supersymmetry turns out to be true, it's relatively easy to explain the low Higgs boson mass. We will look at this in a little more detail next.

Supersymmetry and Naturalness

The unexplained light mass of the Higgs boson is now one of the most pressing issues in particle physics. As you recall from the earlier discussion, the effects due to quantum fluctuations alter radically the mass of the Higgs boson. The observed mass is predicted to be the sum of the theoretical mass and the quantum corrections. The quantum correction term includes two pieces, the maximum energy at which the theory applies and a second term that has in it the effect of the bosons and fermions, with bosons having a positive effect and fermions having a negative effect.

How can supersymmetry tame the quantum corrections? The proper answer is "very elegantly." Remember that the fundamental feature of supersymmetry is that it predicts both fermions and bosons come into the equation in an exactly equal way. This means that for every existing fermion there is a newly predicted boson (and vice versa). Consequently, while the Higgs boson can quantum-fluctuate into a (fermion) top quark, it can also fluctuate into a (boson) stop squark. The fermion adds a negative contribution to the quantum fluctuation, but the boson adds an equal positive contribution. They cancel each other out.

So, we see that the existence of supersymmetry in its simplest form makes the quantum correction to be zero. This is utterly natural if supersymmetry is right. It doesn't take any tuning. It "just happens." This particular consequence of supersymmetry is probably the biggest reason that theoretical physicists like it as a theory. It might be worth noting that, if the mass term in our previous equations really was exactly zero, then the energy term can be arbitrarily huge. In simpler terms, this means that it if supersymmetry actually exists, then a "supersymmetry-enhanced" standard model can apply up to the grand unification energy.

If supersymmetry were perfectly true, the mass of the stop squark would be the same as the top quark (and the same symmetry would be observed for all the other particles), and we'd have seen them already. We should be awash in selectrons just like we are electrons. Since we've not seen them yet, this means that either supersymmetry isn't real or that the supersymmetric cousins are heavier. If the second option is true, it means that the cancellation isn't expected to be perfect. In fact, if the mass of the supersymmetric cousins turns out to be too big, then the cancellation will no longer be good enough to tame the effect of the maximum energy at which the theory applies.

The fact that supersymmetry is such a simple answer to the question of why the mass of the Higgs boson is so small makes it theoretically highly appealing. The fact that the supersymmetric cousins need to be so heavy to account for the fact that we haven't found them yet puts that answer in danger. This tension represents high drama in both theoretical and experimental particle physics. Even if supersymmetry exists, but the cousins are too heavy, then that means that supersymmetry isn't the *only* answer to the mystery of the mass of the Higgs boson (and maybe even not the most important) and that even more new physics is to be found.

In any event, the next time you hear in the news that "supersymmetry is dead," it really doesn't mean that supersymmetry is dead. After all, as we have seen, supersymmetry is a principle and principles are very hard to kill. However, what LHC scientists might have said (and was subsequently simplified by reporters) is that "supersymmetry isn't the explanation for the naturalness problem."

There is another nice feature of supersymmetry related to the Higgs field and the Higgs boson. We mentioned in chapter 6 that the existence of the Higgs field is essentially a theoretical Band-Aid, which is to say it is a patch to the electroweak unification theory. Just like a bandage covers a cut on your finger, the Higgs field fixes the obvious flaw of electroweak unification: the fact that this theory posits that all force-carrying particles had to be massless.

The Higgs boson was incorporated into the standard model in 1967 as an add on. It didn't stem from any deeper physical principles. Theories that incorporate supersymmetry can include the Higgs field in a natural way. And, as we have mentioned here, if supersymmetry is real, the troubling quantum corrections to the mass of the Higgs boson simply disappear. So, on the one hand, supersymmetry addresses some of the unanswered questions of the Higgs field. On the other hand, the mystery of why the Higgs field exists is then transferred to the puzzle of supersymmetry.

Supersymmetry and Experiments

Of course, none of these lovely theoretical ideas are reasons to believe in supersymmetry. You need proof. The one incontrovertible prediction of supersymmetry theory is that there should be twice the number of particles than we currently know. So what do we know of these hypothetical particles?

The answer is not much. Since the data collected so far are adequate to detect low mass objects and we know we haven't found them yet, these hypothetical particles have to be very massive, if they exist at all. This brings us to emphasize an important point. Recall what we're talking about: supersymmetry, where the operative word is symmetry. If the symmetry of supersymmetry was, in fact, symmetric, then the masses of the new particles would be the same as that of regular particles, and we would see that the mass of the selectron was the same as the electron. But we don't. That means that just like the Higgs idea breaks the symmetry between electromagnetism and the weak force, something breaks the supersymmetric symmetry. That's another conundrum, about which we have some ideas. But until we start finding some new particles, this is a concern we can table for the time being.

Given that we haven't discovered these new supersymmetric particles and what we know about the capabilities of our experiments, we can conclude that, if they exist, they must have a mass no lower than about a hundred times heavier than a proton or more. But what would we expect to see if they're real? The LHC is a proton collider and therefore mostly collides quarks and gluons, as they are natural constituents of protons. In a collision making supersymmetric particles, we expect squarks and gluinos to be made most frequently, as these particles also are predicted to feel the strong force.

There are many different types of interactions possible involving supersymmetric particles, but the MSSM makes one useful and pervasive prediction. If a supersymmetric particle is made, then it always has to have a supersymmetric particle in its decay. That means that supersymmetric particles will decay until the daughter particle is the lightest supersymmetric particle, or LSP. Because supersymmetric particles must have a supersymmetric particle as a daughter in the decay and the LSP (by definition) is the lightest supersymmetric particle, there is no lighter possible supersymmetric daughter. Therefore, the LSP is stable. Further, if the universe once made gazillions of supersymmetric particles in the big bang (just as the familiar particles were made), it should contain a similar number of LSPs from all the decays. Given the fact that we don't see them, we know the LSPs must be electrically neutral. Consequently, they can't interact via the electromagnetic force. The strong force is out, too, although the weak force is in according to the theory. So, if the MSSM idea is right, the universe should be full of LSPs—essentially a bath of heavy, neutrino-like things. (It is worth noting that there are supersymmetric theories that don't predict an LSP. In these theories, the supersymmetric cousins can decay into only familiar standard model particles, thereby "erasing" knowledge of their origins. These theories are beyond the scope of this book. Maybe next time.)

Back to the LHC. Let's consider an event in which we make two squarks, like the one shown in figure 7.5. The squarks must be heavy and decay into quarks and LSPs. The quarks make jets as usual, and the LSPs escape undetected, as they don't feel the strong or electromagnetic force. In this particular case, you'd expect to see two jets, and you'd notice that energy is missing. Thus your detector would need to be able to measure jets and also measure energy accurately enough to know some is missing. This is a crucial capability, as the one common

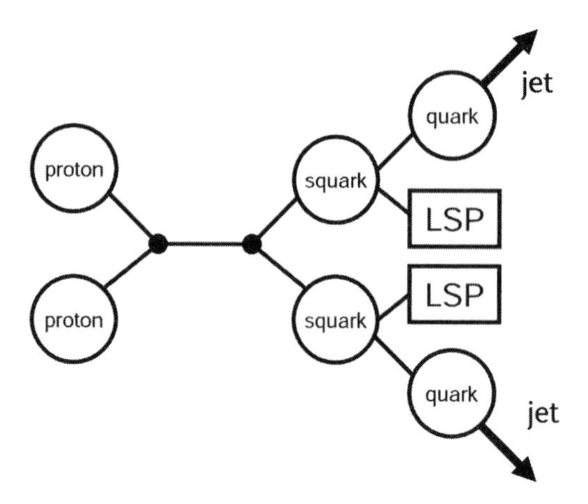

Figure 7.5. Like ordinary particles, supersymmetric particles are expected to decay to particles of lower mass. Supersymmetric particles decay until they end with the lightest supersymmetric particle (LSP) as a daughter.

prediction of essentially all supersymmetry—incorporating theories is an LSP that can escape the detector and thus energy will be missing.

Many types of collisions are possible in the MSSM theory, more than we can discuss here. Even worse, we recall that supersymmetry is a principle and not a theory in its own right, so this means that there are other possible theories that include supersymmetry. Many such theories are much more complex than the relatively simple MSSM. Consequently, killing the idea of supersymmetry will be hard. The best we can do is to kill individual models that incorporate supersymmetry. However, we do know one thing. If (and I stress the "if") supersymmetry is the explanation for why the Higgs boson mass is so low, then we must be able to find supersymmetric particles at the LHC, because if they're too heavy, the cancellation we discussed earlier won't happen. If we don't, then we may not have entirely killed the principle of supersymmetry, but we will have ruled it out as an explanation for the light Higgs boson mass.

Just to make things a bit more interesting, models that incorporate supersymmetry make predictions about the Higgs boson itself. As mentioned earlier, all of the SUSY models actually predict that there is more than one Higgs boson and even that some Higgs bosons carry electric charge. This is clearly a very different prediction than standard Higgs theory. Searches for these sibling Higgs bosons are ongoing.

The LHC and its associated experiments were designed with all these questions in mind. It should be able to resolve some of this theoretical controversy. The next few years running at or near design energy will be unbelievably exciting.

Smaller Than Small

The Victorian era mathematician, Augustus de Morgan wrote:

> Great fleas have little fleas upon their backs to bite 'em,
> And little fleas have lesser fleas, and so ad infinitum.
> And the great fleas themselves, in turn, have greater fleas to go on,
> While these again have greater still, and greater still, and so on.

This oft-quoted passage is a parody of Jonathan Swift's 1733 *On Poetry: A Rhapsody*, which, of course, was written about poetry. However, scientists have taken those lines as a metaphor for the natural world. As one learns about the microworld, one is quickly faced with the observation that all matter is made of molecules. Molecules are in turn made of atoms which are themselves made of electrons and atomic nuclei. The nuclei are made of protons and neutrons and these are composed of quarks. This progression to ever-tinier structures is illustrated in figure 2.4.

However, as far as we know, quarks and electrons are it. That's the end of the line as far as structure goes. Unlike the atom or proton, which have a rich structure with complex interactions between their components, the quarks and electrons are currently believed to have no internal structure at all. Both theoretically and physically, they are considered to be mathematical points.

Of course anyone with an ounce of imagination can't help saying, "Now just hold on a minute. Why couldn't the quarks and leptons themselves have an internal structure?" Well there's only one possible answer and that is "they could." The quarks and electrons (and, by extension, all leptons) could be made of even smaller objects. Or they (rather improbably) may indeed be fundamental (i.e., have no smaller parts, in other words, structureless). In the following pages we consider the evidence for structure as well as how we might out tease out the answer to that question.

Before we proceed further, let's consider the sizes involved. For convenience, they are laid out in Table 7.1. Everything in the microworld is small. A single molecule is so small that you could place a million of them side by side in a single millimeter. They are so small that you can't use ordinary light to see them. And yet, such smaller objects are enormously large: a billion times larger than the research frontier.

Molecules are composed of atoms, which are about a tenth the size of molecules. This factor of ten is not very precise, as there are many kinds of molecules,

Table 7.1. Even the size of the supersmall spans an enormous range

Object	Size (meters)	Size (relative to a proton)
Molecule	10^{-9}	1,000,000
Atoms	10^{-10}	100,000
Nuclei	10^{-14}	10
Protons	10^{-15}	1
Quarks/electrons	$< 10^{-19}$	< 0.0001

from the simple hydrogen, consisting of just two hydrogen atoms (H_2), through simple sugar, with twenty-four atoms ($C_6H_{12}O_6$), to large organic molecules, consisting of hundreds of atoms. However, we can roughly say that a millimeter is 10 million times larger than an atom.

The mental picture of an atom as a little solar system, with the sun as a nucleus and planetary electrons, is flawed and yet it is not without merit. It highlights the fact that an atom consists of mostly empty space, with the electrons swirling frenziedly far from a small, dense nucleus. Figure 2.2 (and the relevant discussion in text near it) gives an idea of the relative sizes involved. Most important for our purposes, it shows just what a tiny fraction of the atom the nucleus takes up. The radius of the nucleus is about 10,000 times smaller than the atom and takes up but a trillionth of the volume.

The nucleus of the atom consists of protons and neutrons, packed tightly together. My mental picture of the nucleus is a mass of frog eggs or marbles after being handled by a toddler with very sticky fingers. Each proton or neutron is about 10^{-15} meters wide, and you would need a trillion laid end-to-end to span a single millimeter. That's small.

Protons and neutrons contain within them quarks and gluons. The simplest way to think of a proton is that there are two up quarks and one down quark stuck in a force field of gluons. Think of three numbered plastic balls in one of those air-blown lottery machines and you get the basic idea.

But the mental picture of quarks as plastic balls has one major flaw. The balls are not much smaller than a lottery machine. Quarks are small. Maybe a better mental picture of the proton is three little flecks of Styrofoam in the same machine.

So what do we know of the size of quarks? Earlier I said that they have no size, and that's certainly how the current theory treats them. However, as an experimenter, I'm more concerned with measurements. You the reader must be curious as to what measurements have revealed the size of a quark to be. And now the answer . . . a drum roll, please . . . they haven't. This doesn't mean we know nothing about their size. We've studied this question rather thoroughly, and we know precisely how good our equipment is. If quarks (and electrons) were larger than about ten thousand times smaller than a proton, we'd have seen that they have a size. In all of our experiments, we've never seen even the slightest believable hint of a size. We therefore conclude that, while we can't say what the size of a quark or electron actually is, we can safely say that if quarks have size at all, they are smaller than one ten-thousandth the size of a proton.

If this idea is hard to understand, let's consider how small an object you can see with your eyes. You can easily see a grain of sand. With very considerable effort, you might be able to see the smallest bit of flour in your cupboard. But that's about it. With your bare eye, you can't see anything smaller. Thus when you decide to look at a germ with your eye, you could conclude that it has no size, but the strictly correct conclusion you should draw is that germs are smaller than a tiny fleck of flour.

With better equipment, say a powerful microscope, one can see that germs actually do have a measurable size. So once you've hit the limitation of your equipment, you simply need to get a more powerful microscope. The microscope that is the LHC and its two primary detectors will observe the size of quarks if they are no less than 20 or 30 thousandths of the size of a proton—or they will set a limit that is about two or three times smaller than currently thought.

While observations, intuition, and de Morgan's ditty may be enough to support a casual suspicion that other levels of matter may occur at ever smaller sizes—a whole new layer or set of layers in the cosmic onion—there are more scientific reasons as well. For instance, consider the periodic table. While Mendeleev intended it to be an organizational scheme, with the formulation of the theory of the nuclear atom and quantum mechanics in the first few decades of the twentieth century, it became clear that the periodic table was actually the first indication of atomic structure, half a century before we truly understood the table's message.

To make this point more clearly, let's focus on the two end columns of the periodic table (see figure 2.1). The leftmost column includes chemically active elements. Hydrogen, lithium, sodium, and all the rest are chemically similar and have the same valences (for those of you who recall your introductory chemistry classes). Yet each of these elements in turn is heavier than the one above it in the column. With our deepening appreciation of chemical structure, we came to understand the increasing mass as being caused by ever more protons and neutrons in the nucleus, while the chemical similarity turned out to be explained by a repeating pattern in the arrangement of atomic electrons, with each of these elements having a single electron available to interact with other atoms. These atoms have differing numbers of electrons, but all but one of them are safely packed away, unable to interact with other atoms.

The story in the right-hand column is essentially identical. Helium, neon, and argon are all chemically similar elements with increasing mass. Chemically, they are all inert because of the arrangements of atomic electrons. These elements also all have their electrons tidily packed away around the atom, with no stray electrons available to interact with other atoms.

While the story told by the periodic table clearly hinted at atomic structure, the story of nuclear radiation also suggested the structure of the nucleus. For instance, cesium ($^{137}_{55}Cs$, with fifty-five protons and eighty-two neutrons) emits an electron and becomes barium ($^{137}_{56}Ba$, with fifty-six protons and eighty-one neutrons). This decay emits a neutrino, too, although that fact was not definitively established until the 1950s. This decay could be understood as having a neutron in the cesium spit out an electron and thereby became a proton. But even before protons, neutrons, and neutrinos were understood, the idea that the nucleus of one element could change into the nucleus of another element was seen as a hint of nuclear structure. Let's take these historical examples and apply the reasoning to the modern world. We realize that historical lessons do not always apply. But sometimes they do.

Our "periodic table" of particles is shown in figure 2.6. Its organization is different from the chemical periodic table. In the figure, there are six types of quarks. The up, charm, and top quarks all have +⅔ charge (in a system where the charge of a proton is +1) and the mass of the charm quark exceeds that of the up quark, which in turn is surpassed by the top quark. Similarly, the down, strange, and bottom quarks all have electric charge −⅓, with the mass increasing as one goes toward the right.

In the case of the leptons, the electron, muon, and tau all have electric charge of −1, with the usual mass pattern. The three neutrinos are all electrically neutral and their mass is not known, although the fact that they have non-zero (and different) mass is not in dispute.

In the modern periodic table, the "chemically similar" units are the rows, in contrast to the columns of Mendeleev's table. We see that there are three "generations" or carbon copies of the same quark and lepton pattern. This is highly reminiscent of the hints that the chemical periodical table was giving us in the latter half of the nineteenth century.

There is another historical similarity to consider. Just like the various atomic nuclei could decay into other nuclei, so too can the quarks and leptons. A top quark can decay into a bottom quark and a W boson. Likewise, the muon can decay into an electron and two neutrinos. These processes are sketched in figure 7.6. Other types of quark and lepton decay are also possible. In fact, all particles in the second and third generations eventually decay into the particles of the first generation. One crucial clue is that the only force that can change one

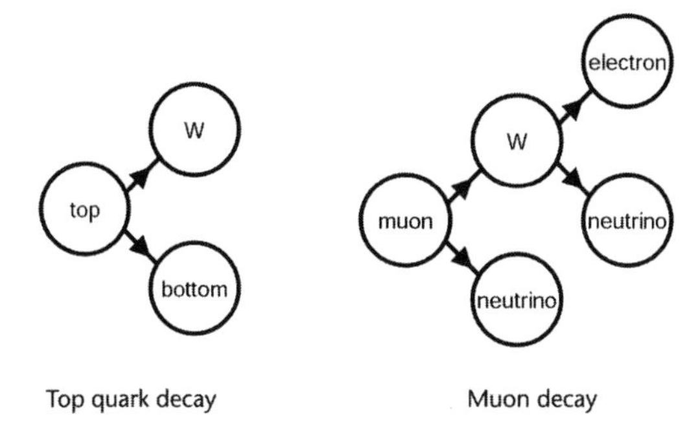

Figure 7.6. The weak force is the only one that can transmute the identity of one subatomic particle into another, shown in these two examples of decay.

quark or lepton into another (we say "change the quark or lepton's 'flavor' ") is the weak force. Further, specifically only the electrically charged W boson can do the job.

There is no hard evidence that the presence of quark and lepton generations indicate that quarks and leptons are themselves composed of smaller (thus far undiscovered) particles. However, the historical analogy is powerfully suggestive and certainly warrants closer attention. The fact that, by emitting a W boson, one can change the quark or lepton flavor is an extremely valuable clue that is screaming something important at physicists.

I just wish that I had the wits to understand what it was saying.

However, even without the crucial insight that cracks the conundrum wide open, we can speculate intelligently on the subject and (much more important) sift through our mounds of data, looking for additional clues. As with all searches for new physical phenomena, you have to make an educated guess about what to look for and then look for it. So, what are the likely experimental signatures of quark or lepton structure? Since the LHC collides protons (which are essentially bags of quarks), we focus on the search for quark structure. Newton's law of gravity treats all objects, even planets, stars, and galaxies, as pointlike particles. As long as you are outside a star, you can replace the entire star with the same amount of mass concentrated at a microscopic point, and you couldn't tell the difference at least as far as gravity is concerned. Once you got inside the star, then the rules would change and the two cases (real star with physical size and pointlike mass) would not equivalent.

More Analogies

Before we get into specifics of quark structure, let's use a few more analogies to get some general concepts under our belts.

What's the Point?

The first analogy we will consider is this: What if a hypothetical being the size of a galaxy had to prove that the Earth had a size and isn't a single point of matter? Consider figure 7.7. One way a galactic-sized being might figure out whether Earth had a size would be to take comets and fling them toward the planet. As long as the comets don't pass closer than 4,000 miles from the center of the Earth (that is, closer than the planet's outside surface), the being can't distinguish between the real Earth and the pointlike Earth. All comets' paths are bent by identical amounts.

However, when the comets are made to pass within 4,000 miles of the Earth's center, well then, the two models give different results. For the pointlike Earth equivalent, Newton's laws still apply. You can use them to calculate how the comet will be affected by gravity. The real Earth will act differently. The comet will plow into the Earth's surface and a new set of physical principles will come into play. The electromagnetic force that governs the behavior of the atoms in the comet and the Earth will determine how big the *splat* will be. So, at a particular size, the relevant laws of physics change, with gravity no longer being the only relevant force and electricity (i.e., atomic forces) taking over.

Bending Waves

The next analogy illustrates how the beam energy matters a lot in the ability to see the supersmall. To illustrate this idea, we need to recall the famous quantum mechanical postulate that objects can act both like particles and waves. It is the wave nature of our particles that is relevant here. For purposes of our discussion, we need to consider two principles: wavelength and diffraction.

Wavelength is the distance between peaks in a wave. The wavelength of a particle is related to the energy of the particle. The higher the energy, the shorter the wavelength, as illustrated in figure 7.8.

Why is the wavelength relevant? Well, this is where diffraction comes in. Suppose you're looking at a lake in which there are waves moving past a stick stuck in the water to measure water level. If the waves are very long, the waves move past the stick without any notice, unaffected by its presence. However, if the waves are short, as shown in figure 7.9, there is a "shadow" of the stick after the waves pass by.

The important points here are the following: (1) seeing something small requires wavelengths even smaller than the object being observed and (2) particles with high energy have a short wavelength. In fact, the wavelength of a particle with the full energy of the LHC is 2×10^{-19} meters, or about ten thousand

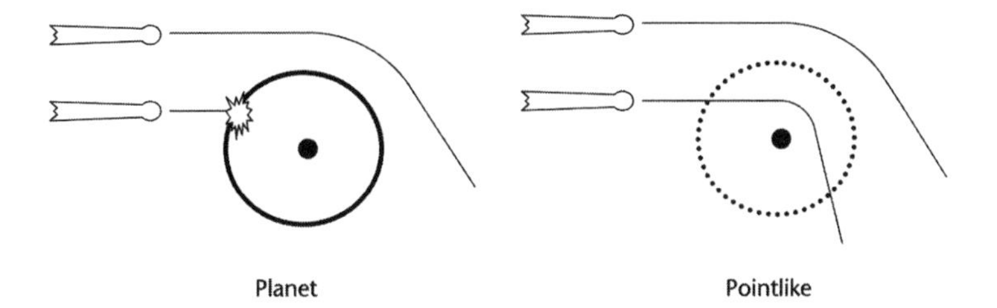

Figure 7.7. As long as a comet's path takes it outside the planet's surface, a planet (*left*) and a pointlike mass (*right*) are indistinguishable from the comet's point of view. However, once the comet's path brings it to a radius smaller than the Earth's surface, the two situations result in different outcomes.

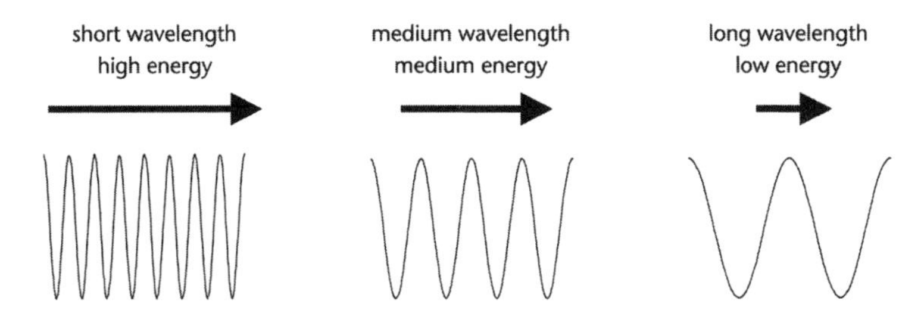

Figure 7.8. Seeing small objects requires short wavelengths and therefore large energies. The effective wavelength is related to the energy of a particle, with lower energy particles having longer wavelengths.

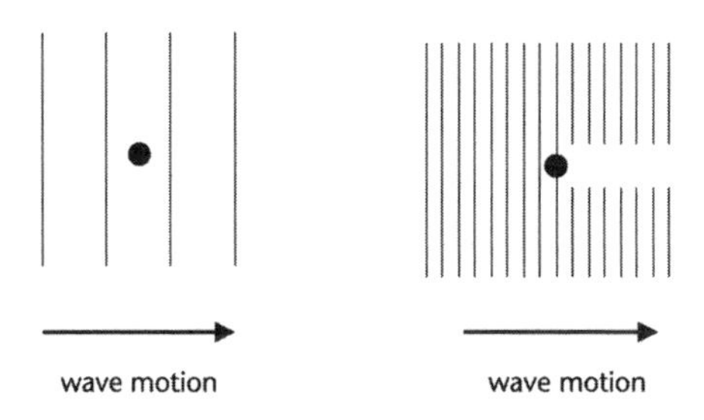

Figure 7.9. When waves pass by an obstacle, such as a stick placed on the water's surface, the pattern of waves can be altered. If the wavelength is larger than the object, the waves pass by unscathed. If the wavelength is smaller than the stick, the waves will be disturbed as they pass by it.

times smaller than a proton. This means that, if the quarks and leptons have a size slightly greater than this number, then the high energy beam of the LHC will be able to distinguish between a pointlike particle and one with a size. By employing some careful analysis techniques, physicists can push their analyses to probe slightly smaller sizes.

Another Layer in the Subatomic Onion

With all these preliminaries out of the way, you might ask: "OK, but what will physicists at the LHC be looking for that could signal quark structure?" There are several techniques that will be used. As with all frontier research, we don't know what the answers will be and therefore we will look in a lot of places. One of them may (and, as usual, I stress the *may*) be the right place to look.

Historically, one of the best places to look has been the most violent collisions. You smash two objects together and see how many collisions there are at each level of violence. Specifically, you look at the amount of "sideward violence." Technically we call this *transverse momentum*, which means perpendicular to the beam. There are technical reasons for this choice, but mostly it is because you have to hit something hard for it to go sideways from its original direction.

Let's look at what an experimental signature of quark structure might look like. We recall from our earlier discussion of the Higgs boson that, if you smash a quark out of a proton, it forms a jet. We can simply add up all the energy of the particles in the jet and that does a pretty good job of looking a lot like the original quark. So we'll just talk about quarks here, although experimentally we measure jets.

In figure 7.10, we see a plot of how often a collision of a particular level of violence occurs. First, we see that low-violence collisions are more likely. Looking at the region of energy labeled "proton-proton regime," that's where the protons are collided with such little energy that the protons act as little billiard balls and there is no hint of the existence of quarks. The dashed continuation of the line shows what the theory predicted would happen if the protons had no structure and always acted pointlike. However, at a particular level of violence, the decreasing trend changed. This is because at that level of energy, the collision became violent enough to see individual quarks, rather than the proton as a whole. Harking back to our analogies in previous chapters, you can think of the protons as finally running into one another or the energy finally being high enough to make short wavelengths. (Indeed, the technical answer includes both ideas.) In any event, the signature that demonstrated the existence of quarks was that at a particular level of violence, there started to be more of

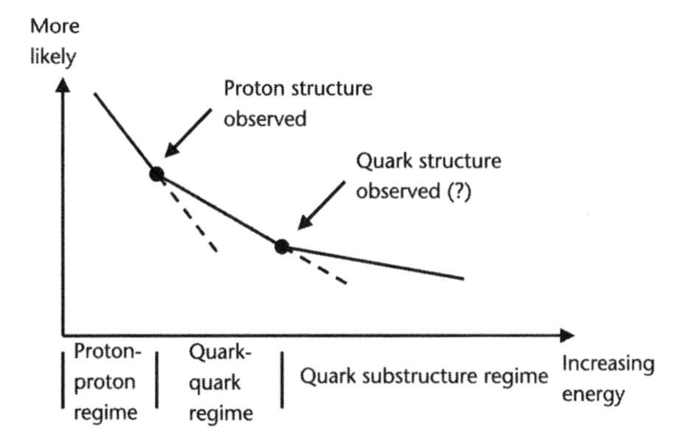

Figure 7.10. The correspondence between the violence (or energy) of a collision and how likely it is. At a low energy, protons collide with such little violence that they don't break up. However, when these collisions become more energetic, the protons begin to fall apart. This allows the quarks contained within them to be seen. With the higher energies of the LHC, we hope to be able to see whether there is a substructure of the quarks themselves. The dashed line indicates the projected behavior protons would show if they did not contain any quarks and the behavior quarks will have if they don't contain anything smaller.

that particular energy collision than you would expect from the lower-energy trend. Quarks were discovered using this and other techniques in the 1970s and early 1980s.

We expect the case to be similar in the event that quark structure is observed at the LHC. Because the energy in the LHC's proton beams is unprecedented, perhaps we will finally make collisions with sufficient violence to start seeing more collisions than the long-observed trend in the scattering of quarks.

There are many ideas as to what might be found inside quarks (including the idea that quarks are indeed pointlike). While the "up-like" quarks have a charge of +⅔, "down-like" −⅓, and leptons −1, if discovered, the objects within quarks could have charges that are a multiple of ⅓, ⅙, or other possibilities. Unlike the case of electroweak symmetry breaking and the Higgs boson, no favorite has emerged from the various contenders. Indeed, with precisely zero direct evidence for the existence of quark and lepton structure, most physicists have taken a "wait and see" attitude, preferring to see what hints the universe will give us. Even so, names have been proposed for these objects smaller than quarks, with the most popular being "preon" (for pre-quark). However, each theoretical physicist who has devised a theory has invented his or her own name, with subquarks, maons, alphons, quinks, rishons, tweedles, helons, haplons, and Y-particles all having been suggested. I kind of like the names quinks or tweedles myself.

Generation Gap

One additional question on the topic of quark structure is the following: Just like the atomic model of the atom has led us to find more and more elements, would it not be true that we would expect additional generations of quarks and leptons? Why are there only three and how do we know there aren't four or more?

The short answer is, of course, there could be more generations. Experiments have tried to find the so-called b-prime and t-prime quarks, which are an unnamed fourth-generation "bottom-like" or "top-like" quark. As this book went to press, no evidence for their existence had yet been found.

Probably the strongest evidence for there being only three generations comes from the LEP experiments that once inhabited the LHC's tunnel. The LEP accelerator collided electrons and positrons in the heart of four superb detectors. Much of the time, researchers tuned the beam energy to make millions of Z bosons. With such a large sample of Z bosons, researchers could study them in great detail. The precision of these measurements is extraordinary. The LEP scientists were able to measure the fraction of time a Z boson decayed into quarks, electrons, muons, and so on. For purposes of our discussion, the interesting decay was when the Z boson decayed into a pair of neutrinos. In figure 6.9, we noted that the Z boson decayed into neutrinos about 20% of the time. Since neutrinos don't interact with matter, these kinds of decays are never observed. However, their existence does make their presence felt. You can calculate how often you expect the LEP beams to make a Z boson depending on how many neutrino generations there are. The LEP experiments concluded that the data were consistent with there being between 2.95 and 3.05 generations. Since the only possible answers for the number of generations are integers (1, 2, 3, 4, . . .), we can state that LEP experiments showed that there were precisely three neutrino generations with considerable certainty.

That's pretty strong evidence that, for whatever reason, the universe allows only three generations. But there is one little bit of wiggle room. Technically, the LEP experiments showed that there were three generations of nearly massless neutrinos. If there is a fourth generation and if the neutrino of this generation has a large mass, then the LEP data can't rule that out. While there is no reason to expect a fourth-generation neutrino to be massive, the fact that the top quark is so much more massive than the other quarks tells us that the idea of a hypothetical heavy neutrino isn't ludicrous. After all, there are ample examples of particles in higher generations being more massive than their lower-generation counterparts, so the idea remains viable. Like anything at the edge of knowledge, only through experiments will the question of quark and lepton structure be resolved.

Preons and Naturalness

How does the idea of preons—as yet undiscovered pre-quarks—and quark/lepton substructure relate to the naturalness problem raised by having a Higgs boson with a low mass? Well, if quarks and leptons are the only particles that are made up of other particles, it doesn't. However, it is possible that the Higgs boson itself could be composite, containing other particles. While we've spoken of preons here, there are many ideas involving composite Higgs bosons, including the idea that they could contain top quarks bound together. This sounds silly, given the mass of the top quarks (173 GeV) and the fact that the bound state would contain two of them (346 GeV). Hold that up to the 125 GeV mass of the Higgs boson and you might say that you could reject the idea out of hand. But, things aren't so easy.

Let's talk about generic hypothetical particles inside the Higgs boson; the discussion here is not specific to quarks. The two (or more) particles inside the boson would have to be held together. This next bit can be hard to get your head around. If the particles are bound together (by some unspecified force), that means it takes energy to pull them apart. Let's consider what that really means. Two particles separated by a large distance each have their mass, which we'll call M. The two of them have a mass of 2M. And, from the magic of Einstein, the energy is $E = 2Mc^2$. Since that c^2 term is a constant, we can drop it in our example. We couldn't do that if this were a serious calculation, but it just simplifies our life a little and we can say that in this case $E = 2M$.

Now we can talk about the mass of the Higgs boson. If the particles inside it are bound (i.e., held together), it takes a force to do that. Prying them apart requires real work (both in the general meaning of the word and in the technical physics meaning). That means, when two particles are near one another, you have to *add* energy to get them apart to counteract the force that holds them together. Conversely, when the particles are bound, they have a lower energy—and, by means of the equals sign in Einstein's equation, less mass—than when they are separated. Therefore, under this scenario, the mass of the Higgs boson can be lower than the mass of its constituents. So, it is possible for a Higgs boson to consist of particles whose individual masses are larger than the Higgs boson.

Getting back to naturalness, recall that the problem arose because the Higgs bosons could fluctuate into other particles, which resulted in some changes in its mass. In the Higgs theory, the change in the mass appeared in the equation as a product of the highest energy possible under the standard model and some sums and differences of the W and Z bosons, the top quark, and the observed mass of the Higgs boson.

We already looked at supersymmetry and how it might solve this problem earlier in the chapter. Having a Higgs boson with structure tames this problematic term in the equation in a different way. While supersymmetry fixed the problem by balancing the sums and differences part of the equation, a composite Higgs fixes it with the other term: the energy involved. The traditional thinking suggests that the Higgs theory should apply up to outrageously high energies. This is the ultimate source of the problem. However, if the Higgs boson has some structure, then the traditional Higgs theory doesn't apply at high energies. At much lower energies, the new physics that binds the Higgs boson together becomes important. This is just like our earlier example comparing a pointlike Earth and a real one and how differently these two planets behave when a comet swings close by. At some size, the two "theories of Earth" behave differently. Similarly, when one gets to an energy at which one can explore the innards of the Higgs boson, the traditional Higgs theory no longer applies.

Of course, there is no evidence that the Higgs boson contains something inside it. In fact, the data thus far are quite consistent with a traditional Higgs boson. But the discovery is fresh, and our measurements are crude. Possibly more precise studies of the Higgs boson will begin to reveal that it isn't the simple particle predicted by Peter Higgs and his contemporaries a half century ago.

Bringing a Whole New Dimension to the Party

There are some things that are so obvious that nobody even thinks to question them. When you got out of bed this morning and dropped by your favorite café for your daily coffee or tea, you moved from one place to the other. This is probably one of times that you say "Well, yeah. Duh . . . " But, if you stop to think about it, there are a lot of things going on in that simple description. It presupposes that you had a location (in your apartment) and you changed that location (to the café). How do we define a location? It needs three bits of information. Your apartment might be located at the intersection of 75th Street and 19th Avenue on the twenty-fourth floor. You could encode that information into simply three numbers (75, 19, 24). The coffee shop might be at 73rd Street, 14th Avenue, ground floor (73, 14, 0). The important point here is the fact that we need three bits of information to locate us in space. That's because we have three dimensions. These dimensions are left/right, forward/backward, and up/down. If you wanted to meet someone at the coffee shop, you'd need to add a time, say 8:30 a.m. With three spatial and one time coordinate, you can completely identify when and where something occurs.

So you knew all that, but you might not have thought through the implications of it. We live in a universe with three spatial and one time dimension. Why is that? Why couldn't we live in a universe with two spatial dimensions, or four? Why three? Well, the short answer is that we don't know. This is one of those currently unanswered questions. But unanswered questions require answers, which may in turn beget more questions. So the first question is "Is it absurd to ask if there are other dimensions, given that our experience so clearly demonstrates that we live in a three-dimensional reality?" The answer to that question is both yes and no. Obviously if there are other dimensions, they are somehow different from the familiar three.

In a moment, I will describe some of the reasons that the seemingly absurd hypothesis that there are more spatial dimensions than the familiar three is an interesting idea. For the moment, let's consider what kinds of extra dimensions could exist. We already know that another dimension like left/right or up/down is impossible. If it (they?) existed, we'd have seen them already. (There is one caveat. That statement only really applies to the four known forces. If there were additional dimensions into which our familiar particles are somehow forbidden to move, these could still exist. Since we're trying to extend our understanding of familiar matter, we ignore this possibility for the following discussion.) However, there is a class of extra dimensions that could exist that haven't been ruled out yet. These extra dimensions are smaller than the familiar three.

For dimensions to be smaller, this implies that they must be cyclical. That's a complex thought, so let's spend a little time on it. The familiar left/right dimension is infinite. In principle, you can go left or right for an infinite amount of time. There are no limitations on the distance you can travel. If we invented a new dimension that had the same basic property (unlimited extent) but was somehow "smaller," what does that mean? If the new dimension was half the size of the familiar kind or a tenth or even a hundred-thousandth, you're still talking a huge size. A tenth of infinity is still infinity. While such a dimension might look a bit different compared with our familiar three, such "infinite but smaller" dimensions are clearly ruled out.

Cyclical extra dimensions are not ruled out. Let's consider a simpler case. Rather than discussing our familiar three infinite dimensions and adding another smaller and cyclical dimension, let's instead talk about a world with a single infinite dimension and then add a smaller cyclical one. I am talking about, of course, tightrope walking.

Tightrope walkers (well, successful ones anyway) live a constrained existence. They can walk forward or backward. They can't move left or right. They can't move up or down (beyond falling and that makes them unsuccessful). They live in a strict one-dimensional world.

In contrast, ants are much smaller and have more options. While they can

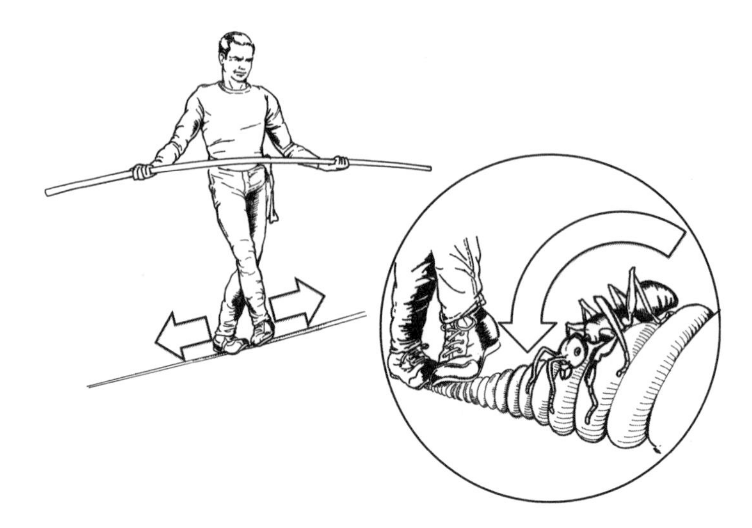

Figure 7.11. A tightrope walker can move only in one dimension (forward/backward) on the rope. In contrast, an ant, being much smaller, has access to two dimensions, both the one accessible to the human and another, smaller one. This smaller dimension is also different in that it is cyclical, meaning that the ant can walk around the rope and end up back where it started. Figure courtesy of Dan Claes.

walk along the rope in the way the tightrope walker does, they can also walk around the rope. This second dimension is quite different, since, if they walk around the rope, they quickly end up back where they started. This is what is meant by "cyclical." These ideas are illustrated in figure 7.11.

Further, the tightrope walker is totally unaware of the extra dimension that exists under every one of his steps. He is simply too big to be able to avail himself of the benefits of those extra dimensions. If they could, then two tightrope walkers on the same rope could pass one another, just like two ants could. We see that being smaller allows someone to exploit the extra dimension, while a larger entity cannot.

We can come up with other analogies. Instead of the traditional cable used by tightrope walkers, suppose that instead they were walking along a string of pearls. The walker is again restricted to one dimension, while a creature smaller than an ant . . . say a flea or a microbe . . . could walk around on the surface of each pearl, just as we can walk on the surface of the Earth. These much smaller creatures could access not one but two additional small and cyclical dimensions.

The number of additional dimensions is not limited to one or two; in fact there could be very many additional small dimensions. However, to generalize to five or six or more extra dimensions is hard on the imagination, so you'll have to just trust me that mathematicians are quite clever and have managed to extend their theories to an arbitrary number of dimensions.

We can kind of visualize a three-dimensional space with two small extra dimensions. Imagine a room full of tiny, tiny BBs, with each BB being smaller than an atom. You could specify your location using the ordinary three dimensions to identify which BB you were at; then for each BB you could use the two smaller dimensions to specify where you were on the surface of the BB. This is an instance of three infinite dimensions and two smaller and cyclical ones. (A purist might—correctly!—point out that this isn't a perfect analogy. As you move around the surface of the BB, you're actually moving in the familiar three-dimensional space. However, if you use the three dimensions only to identify the BB, it's a reasonable analogy if it's not pushed too far.)

So all the discussion thus far on extra dimensions is rather abstract. What good is it, physically speaking? Many physical problems can be explained if the hypothesis of extra dimensions were true. One such idea involves a hypothesis called *superstrings*. Superstrings are not a subject on which the LHC is expected to shed much light, so we do not explore them in detail in this book. Briefly, the hypothesis is that quarks and leptons are not the ultimate building blocks of the cosmos, neither are any pointlike particles. Instead, the smallest objects possible are called *strings*, which can be envisioned as little sticks of spaghetti or miniscule hula hoops. These strings can vibrate, with the different notes being different particles. A C-flat might be an electron, while a G-sharp might be a top quark. The theory is, of course, far more complicated than that, but this gives you the basic idea.

When theoretical physicists started playing with this idea, trying to see if it made sense, they would try to calculate things like what happens when two strings interact. Each kind of interaction was given a percentage probability—an 8% chance to do this, a 15% chance to do that. However, when all the percentages were added up, the result wasn't 100% but instead a really huge number. Anything other than 100% is absurd, so that's enough to kill the theory. But theorists are clever lads and lasses and like to play with ideas, so someone said, "Well what if we redo the calculation with four space dimensions?" When they did the calculation, they still found that the sum of probabilities exceeded 100%, but the sum was smaller and more reasonable. Extra dimensions were added one by one, and, when ten dimensions were used, everything fell into place and a result of 100% was obtained. (A more sophisticated calculation later bumped the number of dimensions to eleven.)

The extra dimensions beyond the familiar three of space and one of time were all itty-bitty additional dimensions. But it is a clear requirement of super-string theory that additional dimensions exist. This is hardly reason to believe that these extra dimensions exist, but is certainly a reason to be interested in the idea.

Another nifty thing one can do with extra dimensions is explain why gravity is so much weaker than the other forces. Recall that if we use the strong force to set the standard of force strength, the electromagnetic force is a hundred to a thousand times weaker, while the weak nuclear force is about 100,000 times weaker. However, gravity is crazy weaker, in fact, some 10^{40} times weaker. One idea invented to explain this huge hierarchy in the forces is that there are additional smaller dimensions into which gravity can spread but that other forces cannot. Under this thinking, the electromagnetic, weak, and strong forces spread out into three space dimensions, while gravity might spread out into more. Considerable effort has been put into this idea, as well as many experimental tests. Thus far, there is no indication that the conjecture is true. It is perhaps worth noting that the silly hysteria of 2008—when people were worried about the LHC making black holes—was rooted in this idea. Essentially what was claimed was that, if the LHC could see small enough sizes to look at tiny extra dimensions, then possibly this would mean that gravity would suddenly become as strong as the other forces. If that occurred, then subatomic black holes would be created and the Earth would get swallowed up. While it was a worthwhile question to ask, we know that there is exactly zero danger of its occurring. The Earth is constantly pelted with cosmic rays from space with energies incomparably higher than the LHC. This has been going on since the Earth was formed, some 4.5 billion years ago. If the LHC could make black holes, then black holes would have been created countless times in the atmosphere of the Earth between these hyperenergetic cosmic rays and the molecules of the air surrounding our planet. We're still here, ergo, there was never any danger. While the black hole hysteria was never very serious from a scientific point of view, it's worth remembering that for it to be even theoretically credible required these improbable extra dimensions.

Additional extra dimensions also can help with the naturalness problem surrounding the Higgs boson. Recall that quantum mechanical effects caused by the Higgs boson fluctuating into other particles predicted a modification to the simple and obvious mass of the Higgs boson. This correction consisted of a term that involved the maximum energy at which the standard model applied multiplied by another term with sums and differences of the masses of the particles into which the Higgs boson temporarily converted.

Extra dimensions can tame the naturalness problem by limiting the maximum energy at which the standard model applies. It makes sense if you think about it. The standard model was constructed within the known universe with its three—or four—dimensions. If, at a particular energy, we detect other dimensions, the standard model will have to be modified. Since the theory will no longer apply, the maximum energy that the Higgs theory applies will be much smaller than the much higher Planck energy. In this way, the naturalness problem could be solved.

We should always keep in mind that the idea of extra dimensions is wholly speculative. There is exactly zero experimental evidence that they exist. However, neither has the idea of extra dimensions been ruled out. So, we keep the idea around and continue to test it. During the running period of 2010 to 2012, the LHC set limits on the size of extra dimensions, which means that if they exist, they must be truly tiny indeed.

What's the Matter with Antimatter?

"Space, the final frontier . . ." is the opening line of a wonderful television show from the 1960s called *Star Trek*. A youthful Don would look forward to watching it in syndication, peering at a fuzzy picture on a UHF station. (You youngsters ask your parents what UHF was. If you know, don't admit it, because that means you're getting old.) In this show, a mighty starship called the U.S.S. *Enterprise*, captained by the legendary James T. Kirk, would scoot around the galaxy encountering situations that frequently had moral relevance to the social problems of the day.

For our purposes, the show itself isn't as important as the ship's engines. They were powered by antimatter. It could be true that, like many of the high-tech doodads in the *Star Trek* universe, antimatter was merely a fictional device, on par with dilithium crystals; a convenient futuristic plot device to make plausible their speedy journeys.

Unlike many things that appear in science fiction, antimatter is entirely real. (Further, antimatter could be used in principle as a spacecraft propellant, as it is the highest energy power source ever discovered, making not so silly its presence in a starship's engines.) Antimatter is the opposite of matter and, when combined with matter, will completely annihilate into pure energy.

Antimatter is perhaps most simply understood at a particle level. For every particle discovered, there is a corresponding antiparticle. There are antimatter electrons (called positrons) and antiquarks, which have no special name. Some particles, like the photon, are their own antiparticle. From antiquarks, you can make antiprotons and antineutrons. Toss in antielectrons and you can make anti-atoms. With anti-atoms, you could in principle make anti-anything: anti-you, anti-me, antipasto (or should that be anti-antipasto?), and so on.

Antimatter can be created in physics laboratories by converting a prodigious amount of energy into matter and antimatter. In fact, it would take the entire

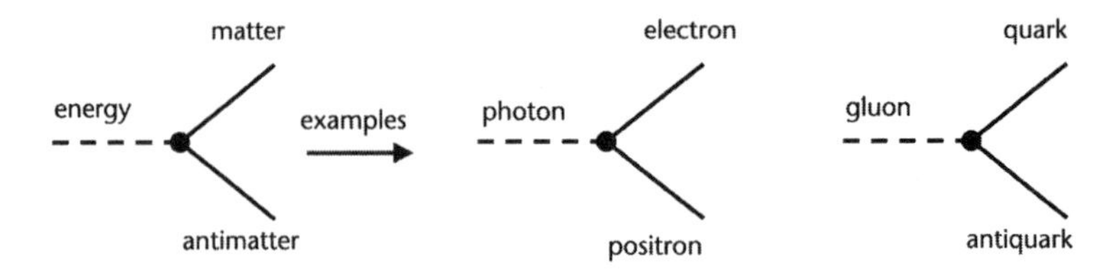

Figure 7.12. Energy or force particles can be turned into antimatter but only with a corresponding matter particle.

energy of a Hiroshima-type nuclear explosion, converted with 100% efficiency, to make enough antimatter to make an anti–paper clip (see figure 2.8).

The operative words here are "into matter *and* antimatter." The way antimatter is made is to convert energy into *identical pairs* of matter and antimatter (figure 7.12). Yet this leads us to one nagging problem.

If we combine two observations: (1) matter and antimatter are made in equal pairs and (2) we only see matter in the universe, we're led to the obvious question, "Where the heck is all the antimatter that should be here?" This remains one of the unsolved mysteries of science.

We do know some things about how the universe treats matter and antimatter. To the best of our knowledge, the strong and electromagnetic forces treat matter and antimatter identically. However, the weak force doesn't. The entire story of that discovery is rather complex, starting in 1956, with experiments involving decay in the nucleus of cobalt atoms. In these decays, matter and antimatter acted differently. To make up for this, scientists dragged in the nature of space and said that the equations governing an antimatter particle moving to the left were equivalent to a matter particle moving to the right. (The actual directions are arbitrary; what is important is that they were the opposite.) In 1964, another paradigm-shifting experiment revealed a slight asymmetry in the decay of neutral K mesons, or kaons. Neutral kaons are hadrons, which means they contain quarks and experience the strong force. A meson is a particle containing one quark and one antiquark. Neutral kaons contain either a strange quark and a down antiquark or a down quark and a strange antiquark. Fitting the asymmetry into the standard model was solved by bringing in the fourth dimension (i.e., time), and now the equations are such that antimatter/left/backward in time is identical to matter/right/forward in time.

In 1999, additional measurements involving kaons revealed more about the matter-antimatter asymmetry. While learning that some physical processes favor matter over antimatter was a huge step in understanding why we live in a matter-dominated universe, the simple fact was that the slight preference for matter observed in neutral kaon decays wasn't enough. There had to be more.

Shortly after the discovery of the bottom quark in 1977, physicists calculated that they expected a greater asymmetry in mesons involving bottom quarks. The story of the study of neutral kaons should not be treated this cursorily, because it involves brilliant scientific detective work. But to limit the scope of this book and to focus on the LHC, the study of neutral kaons is only sketched here. The full story is given in the suggested reading.

In 1999, two experiments—Belle in Japan and BaBar in California—turned on with the sole purpose of making bottom-quark-containing mesons (B mesons) in prodigious quantities, with neutrally charged ones being of particular interest here. While researchers at these experiments had intermediate successes, in 2004 they announced an enormous asymmetry for matter over antimatter in the decay of neutral B mesons. This preference for matter is a hundred thousand times greater than that seen in the decay of neutral kaons. Because the study of hadrons containing bottom quarks is an important goal for the LHC, we will sketch the Belle and BaBar method here.

Both Belle and BaBar were detectors designed to study collisions of electrons and positrons. And not just any old collisions; the beam energies were carefully selected to produce the $\Upsilon(4S)$ meson (that's the capital Greek letter upsilon four S), which consists of a bottom and antibottom quark. The $\Upsilon(4S)$ can decay into two neutral B mesons. These daughter mesons are called $\bar{B}^0$, consisting of a down antiquark and a bottom quark and B^0, consisting of a bottom antiquark and a down quark. You'll note that this is similar to the neutral kaon case, with the bottom quark taking the place of the strange quark. Figure 7.13 shows how the $\Upsilon(4S)$ can decay, highlighting that matter bottom quarks (denoted b) and antimatter bottom quarks (denoted $\bar{b}$) occur in equal quantities.

In order to study the decays of these neutral B mesons, a particular mode was used. This decay was into a charged K meson and a charged π meson, also called a pion. Because the B meson was electrically neutral, the K and π mesons had to have opposite electric charges to balance out. These mesons consisted of the following quarks: K^+ (up and anti-strange), K^- (strange and anti-up), π^+ (up and anti-down), and π^- (down and anti-up). The need for this level of detail will become apparent in a moment.

Figure 7.14 shows how B^0 and $\bar{B}^0$ mesons decay. Most readers can ignore everything between the circles on the ends. The most important thing to note is that if you see a $K^+ \pi^-$ decay, you know it came from a B^0. A $K^- \pi^+$ decay comes from $\bar{B}^0$.

So, here's the big point. Because B^0 and $\bar{B}^0$ mesons are made in equal quantities, you'd expect to see the $K^+ \pi^-$ and $K^- \pi^+$ decay modes occur with equal frequency. But both Belle and BaBar didn't. Both saw that about 80% of the time they decayed to K^+ compared with 20% for K^-. That means the number

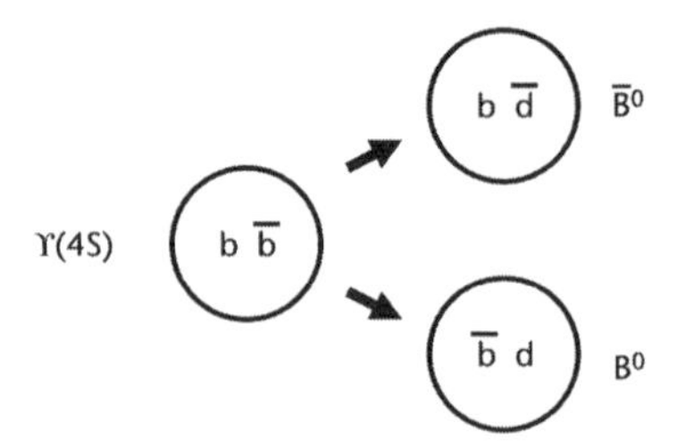

Figure 7.13. Of the many ways in which the Υ(4S) can decay, an especially interesting process for studying matter/antimatter asymmetry is the decay into pairs of neutral mesons containing bottom quarks.

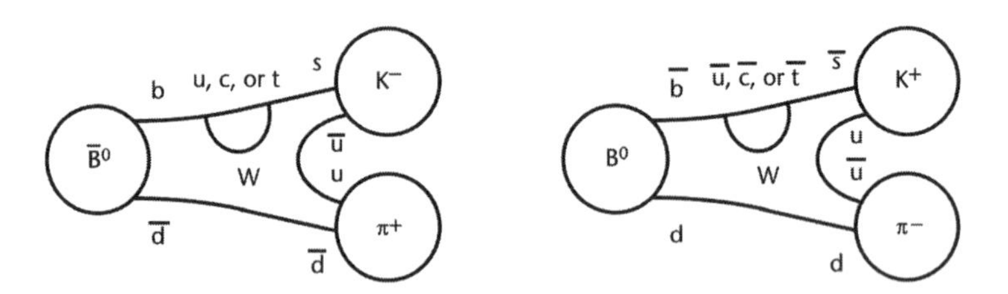

Figure 7.14. Once these mesons containing bottom quarks are made, studies of their subsequent decays are an excellent probe of matter/antimatter asymmetry. The decay of those mesons containing bottom quarks are shown here.

of antimatter bottom quarks seen was about 80% that of bottom quarks: less antimatter than matter.

This measurement firmly established that the study of bottom quarks could easily be crucial for shedding light on the preponderance of matter in the universe. In both Belle and BaBar, these measurements took about 200 million Υ(4S) and ended up with about a thousand charged K and π decays. A thousand samples of the desired decay aren't all that many, highlighting the need for more data. In addition, the decay mode described here is only one of many being investigated. Thus many more examples of bottom-quark-containing hadrons can only help the situation.

The LHC is, in many ways, a superior source of bottom quarks compared with Belle and BaBar's electron/positron collider. Belle and BaBar's strength was the fact that they had an exquisitely pure sample of specific bottom-quark-containing hadrons. Unfortunately, this is also their weakness. They can make what they make and that's it. In contrast, quarks abound in the LHC. Many different hadrons containing bottom quarks are possible, allowing for a much richer set of studies. These studies are a crucial point of the LHC's design and research goal. While LHCb specializes in investigating bottom quarks, ATLAS and CMS are also serious players.

Heavy Ions

Although most of the LHC's experimental program focuses on colliding protons together at the highest energy, it is not everything the LHC does. About one month each year, the LHC has and will continue to accelerate the nuclei of lead atoms and collide them together. The physics questions being explored in lead-lead collisions are vastly different than those in proton-proton ones. In proton collisions, the idea is to focus as much energy in as tiny a spot as possible, like the pressure at the point of a pin. In the case of lead collisions, the idea is to spread a lot of energy over a large volume. Energy across a large volume can reveal phenomena that a focused-energy collision would miss.

Like all other facets of the LHC experimental program, there are countless different questions one can ask and measurements one can make while colliding lead nuclei. However, one phenomenon stands out in the study of lead and indeed all heavy nuclei collisions: observing and characterizing an entirely different state of matter.

People are familiar with the three most common states of matter: solid, liquid, and gas. When you think about it, the insight that the same materials can have such vastly different properties and yet still be the same thing is pretty amazing. Air and a frozen clod of dirt (i.e., a gas and a cold solid) are entirely different things and yet steam and ice (also a gas and a cold solid) actually *are* the same thing.

We call these various states of matter *phases*. What most people don't know is that there are other phases, both observed and merely hypothesized. Typically one can change matter from one form to another by heating or cooling it (or equivalently adding or subtracting energy). Let's think about what happens to water when you add energy to it. Start with a familiar ice cube. If you heat it, first the ice cube warms up. When it reaches 32°F (0°C), the ice melts. Water changes from its solid form to its liquid form, that is to say it changes phase. Heating the water changes it to steam, water's gaseous form. In its gaseous form, individual water molecules can fly around, interacting very little with one another. In contrast, water in its liquid form exhibits very different behavior. Molecules of liquid water "know about" each other. That's why liquid water can experience such behaviors as viscosity and surface tension. The fact that the same matter can act so differently under different energy and temperature conditions is one of the reasons we study it. We want to see all the rich behaviors that matter can exhibit.

Since gaseous water consists of individual molecules, we need to know about water molecules. Water molecules consist of three atoms, two of hydrogen and one of oxygen. (Hence, water's molecular formula: H_2O.) In its gaseous form just above the temperature at which water boils (212°F or 100°C), each molecule acts individually.

However, as the temperature of the steam is increased, the water molecules bounce around with more and more energy. Eventually they start bouncing into one another hard enough that the molecules are broken apart, with hydrogen and oxygen atoms individually wandering around willy-nilly.

While the electrons are strongly bound to their respective nuclei, as the temperature is raised further, the nuclei are no longer able to hold onto their electrons, and they are stripped away. Oxygen and hydrogen nuclei are intermixed with free electrons. The whole mix is electrically neutral. This is actually considered a new stage of matter called a *plasma*. You can see an example of an electrically produced plasma in a fluorescent light bulb or in a plasma television. A particularly cool example of a plasma is one of those "Eye of the Storm" plasma globes.

Further heating this mixture eventually will cause nuclei to break apart, leaving electrons, protons, and neutrons flying around. Temperatures as hot and energies as high as this have been achievable for decades.

We recall that protons and neutrons are made of quarks. Each proton could be thought of as roughly a bag with three quarks locked firmly within it. Technically, as was discussed in chapter 2, we say that the quarks are "confined" in the proton or neutron. The question to be asked is whether at high enough temperatures the quarks can be freed from their protective nucleonic cocoons?

Above a certain temperature, the quarks are deconfined and allowed to mix freely (figure 7.15). In some ways, the situation is analogous to a tumbler full of ice cubes (the protons and neutrons in the analogy), which melt when heat/energy is added to form liquid water (the intermixed quarks). This state of matter has historically been called *quark-gluon plasma* in analogy with the electrical plasma of fluorescent lights.

The ice cube analogy is more apt than you might think. For a long time, it was thought that a quark-gluon plasma would behave like a super-hot gas, with the quarks bouncing around, ignoring one another. However in 2005, experiments at the Relativistic Heavy Ion Collider, or RHIC, on Long Island showed that, when nuclei of gold collide, what results is a new form of matter that acts more like a liquid. In fact, when nuclei are heated to the point where the protons and neutrons melt in place, the freed quarks and gluons act like a liquid with zero viscosity.

Viscosity is a property of liquids that relates to how thick they are and how

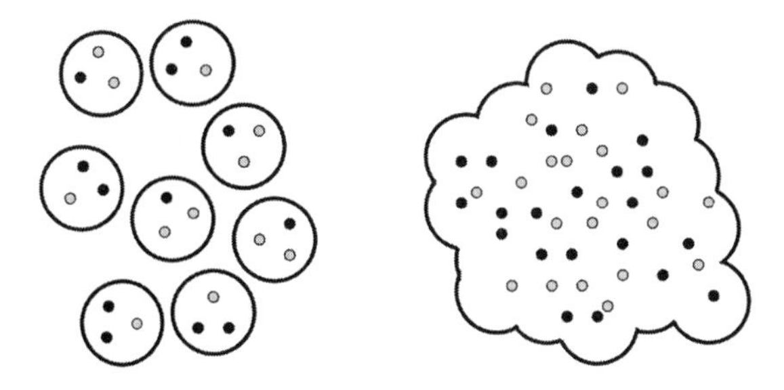

Figure 7.15. In ordinary matter (*left*), quarks are held inside protons and neutrons, three at a time. In a quark-gluon plasma (*right*), the quarks are no longer held inside the nucleons and are allowed to intermix freely.

much they slosh. Take out a spoon, stir your coffee, and remove the spoon. The coffee will continue to swirl around in the cup. Do the same things in a bowl full of warm honey and the swirling will die down quickly. We say that honey is more viscous than water. Rather surprisingly, when quarks were freed from the protons and neutrons, the resultant state of matter acted like a fluid that swirled forever. It had zero viscosity.

Just precisely how does one free the quarks? By heating up a nucleus, of course. Specifically, one aims beams of atomic nuclei at one another. These nuclei generally miss one another or occasionally experience a grazing impact. However, once in a great while, the two nuclei hit head-on. Just like hitting two bullets together at high speed can cause them to melt, so do the nuclei. (If you have trouble believing that an impact can make something warm, try banging a hammer many times on a solid rock and then feel the head of the hammer. It will be hot.)

Figure 7.16 shows an example of such a collision. In figure 7.16a, the two nuclei are coming together at high energy. While nuclei are basically spherical, in a particle accelerator, they look more like two pancakes hitting face on. This is because of Einstein's theory of special relativity, which says that fast-moving objects will contract, but only in their direction of motion. Thus the sideways dimensions are not shrunken and remain circular. The result is this pancake.

In figure 7.16b, the pancakes pass through one another, with some of the energy being deposited in the nuclei, heating them up. If the conditions (i.e., energy) are right, the nuclear matter will be heated enough to free the quarks and a quark-gluon plasma will be formed, as seen in figure 7.16c. Finally, as 7.16d shows, the ensuing fireball will continue to expand and cool off, with the quarks recoalescing into protons, neutrons, and other hadrons. The collision is over.

A fair question one might ask is this: "How would you know a quark-gluon plasma if you saw one?" Like every question asked at the LHC (or any modern

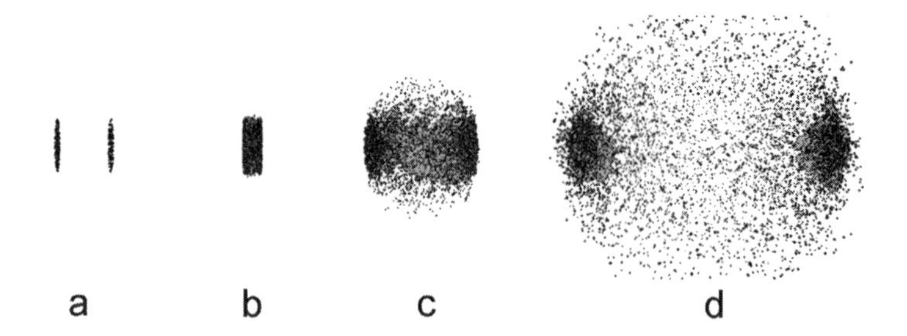

Figure 7.16. Stages in the formation of a quark-gluon plasma: *a*, Two nuclei approach each other; *b*, the collision of the two particles; *c*, the quark-gluon plasma forms as the shock of the collision heats the nuclear matter; *d*, eventual expansion and cooling of the fireball. Courtesy Jeffrey T. Mitchell, Brookhaven National Laboratory; simulation by the UrQMD Collaboration.

experiment), there will be many different ways to answer. But we can discuss one way that will illustrate the important points. This is a technique called *jet quenching*.

Jet quenching is a pretty cool and intuitive idea. Earlier, in our discussion in chapter 6 of the experimental signatures of the Higgs boson, we mentioned that if you knock a quark out of a proton or neutron, a jet will form. Jets, as you might recall, are "shotgun blasts" of particles that are the characteristic signature of a quark escaping a proton.

In a collision between any heavy nuclei (e.g., lead or gold), two quarks will sometimes hit one another just like when beams consisting only of protons collide. In the case of a collision that is violent enough to dislodge quarks from a nucleon but not violent enough to form a quark-gluon plasma, the scattered quark can escape the fireball mostly unscathed. The quark passes by the protons, neutrons, and other hadrons in the fireball. Because these hadrons have no net color (the strong force discussed in chapter 2), effectively the quark simply doesn't "see" them. This behavior is illustrated in figure 7.17.

However, if the fireball is hot enough to melt the protons and neutrons into a quark-gluon plasma, then the violently scattered quarks will pass by free quarks, each with its own color. These high-energy quarks will then bounce into the quark-gluon plasma quarks and sometimes never make it out of the fireball. So you'd expect to see that as the collision becomes hotter and hotter, you'd see fewer and fewer high-energy scattered quarks (and therefore fewer jets). Thus we say that jets will be "quenched." This is one of the many signatures that physicists look for in the LHC experiments. Tentative evidence for jet quenching was seen at RHIC and definitively observed at the LHC by the ATLAS collaboration in 2010, followed closely thereafter by the CMS collaboration.

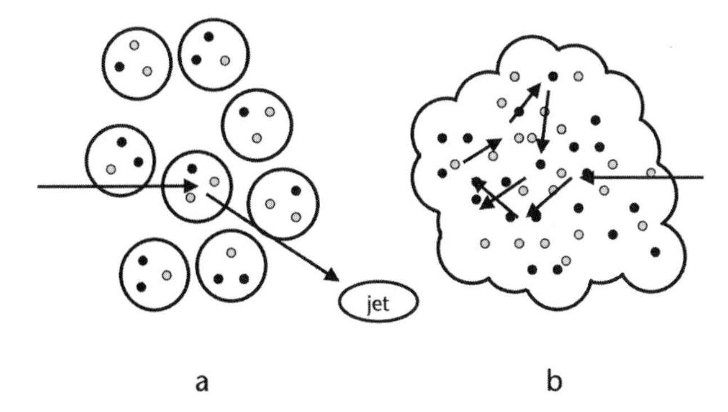

Figure 7.17. Jet quenching. ***a,*** **In ordinary nuclear matter, a quark will scatter and leave the volume, making a jet.** ***b,*** **In a quark-gluon plasma, the quarks will bounce around, hitting the free quarks. Thus the quark does not leave the volume as often, with fewer jets being produced as a result.**

Why is this kind of study interesting? It's because collisions like these recreate the conditions of the universe just instants after its creation. This is not a book on cosmology, but briefly we know the universe was once much smaller and hotter. This is a prime element of the big bang theory. Much of the details of the very early universe are unknown, and only through observations of matter in particle colliders, paired with a dash of intelligent speculation that we can guess what might have occurred.

We believe that the universe once had a temperature that was high enough that the weak force and the electromagnetic force acted the same. Above this temperature, we say that the electromagnetic and weak forces were symmetric. The study of the breaking this symmetry—called *electroweak symmetry breaking*—is one of the important goals of the LHC. No matter the mechanism that breaks this symmetry, be it the Higgs idea that is strongly supported by LHC data or something else, it is thought that the universe cooled enough that the electromagnetic and weak forces became two distinct phenomena by about a trillionth of a second after the big bang.

For the period of about one trillionth of a second until a millionth (10^{-12} to 10^{-6}) of a second, the quark-gluon plasma is thought to have reigned supreme. The entire universe was so hot that quarks could (at least in principle) swim from one side of the universe to the other, unfettered by such considerations as protons and neutrons. It is this period in the history of the universe, called the "quark epoch," that the study of heavy ions is intended to illuminate.

At the end of the first millionth of a second in the history of the universe, matter had cooled enough that quarks and gluons could not move around at will. Just like water freezes at a certain temperature, the quark-gluon plasma froze out, leaving the quarks firmly ensconced in the resultant protons and neutrons. The universe would eventually cool further, allowing protons and neutrons to combine to make helium nuclei (consisting of two protons and two neutrons). Further cooling would let electrons attach to nuclei to make hydrogen and helium, which in turn would slowly coalesce into stars and galaxies.

But all the physical phenomena that govern that later cooling are relatively well known. It is the phase transition between quarks trapped in protons and neutrons and the free-ranging, low-viscosity liquid quark-gluon plasma that heavy ion collisions at the LHC have begun to explore with some first tentative steps. Perhaps the LHC will attain temperatures that could reach another phase transition from the low viscosity quark-gluon plasma to something more akin to a gas. Only time (and experiments) will tell.

Other Questions

The kinds of topics discussed thus far all tend to cluster near the very edge of our understanding of the universe and, for all of them, the LHC may reveal phenomena never before observed. In addition to frontier-blazing experiments, there are questions that scientists ask that are more evolutionary, that is to say simple improvements and extensions of our understanding of phenomena about which we already know a great deal. In fact, all of the four large detectors at the LHC (ALICE, ATLAS, CMS, and LHCb) described in chapter 4 will spend a great deal of their time on just these kinds of measurements.

However, two phenomena are beginning to be studied and will continue to be studied at the LHC by small, dedicated experiments, called TOTEM and LHCf (for LHC forward). While the thrust of this book is the new horizons, the uncharted vistas that we hope the LHC will let us discover, these measurements that lead only to an incremental understanding of the universe are interesting too, and we will mention them briefly here. These two phenomena are called *proton diffraction* and *cosmic rays*. Let's look briefly at both of them.

Proton diffraction takes its name from an optical analogy. In optics, diffraction is the phenomenon whereby light waves can bend around corners. In fact, diffraction is a phenomenon exhibited by all waves, as shown in figure 7.9 in our discussion of quark structure searches. Recall that all particles have a wave equivalent, and the protons in the LHC are no exception. Thus, it is expected that protons will exhibit diffractive behavior when they pass by one another.

Indeed, this kind of interaction between protons has been observed and studied for decades, until recently at Fermilab's Tevatron in which protons and antiprotons were accelerated to an appreciable fraction (14%) of the LHC's design energy. So studying this phenomenon at the LHC is expected to extend

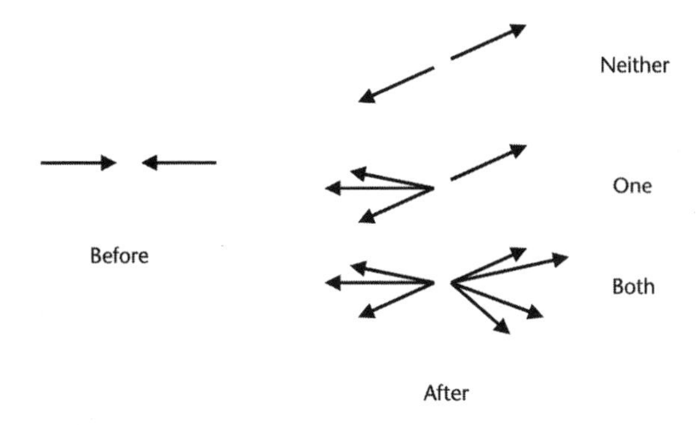

Figure 7.18. When protons collide, one, both, or neither proton might break up.

our current understanding, rather than open up an entirely new realm to study. Of course, there can always be surprises. This is the research frontier, after all.

The experimental signature for these kinds of studies is quite distinct. When the two protons collide, one or both of them survive the collision intact. One can contrast these kinds of collisions with the ones that are most frequently studied at the LHC, ones in which both protons are torn completely apart. Figure 7.18 illustrates the differences. The destructive collisions are the most studied at the LHC, as they are the most violent and are most likely to reveal new physical phenomena. However, they are quite rare. Collisions in which at least one proton survives intact form the vast majority of proton-proton collisions. The TOTEM experiment is designed to explore this well-studied phenomenon at the higher energy the LHC is providing.

Cosmic Rays

Cosmic rays are a generic term for particles that rain down on the Earth from outer space. They were discovered well over a century ago, when physicists used the Eiffel Tower and the then-newfangled hot air balloon to show that air was more conductive at great altitudes than it was on the ground.

Our understanding of cosmic rays has improved dramatically over the past hundred years. We now know that highly energetic particles from space, typically protons, hit the Earth's atmosphere and slam into air molecules high above the Earth's surface (figure 7.19). The most interesting cosmic rays are highly energetic, indeed much, much higher than any particle beam we can make on Earth. To give some perspective, the highest energy cosmic rays have about a hundred million times more energy than the beams at the LHC.

Studying cosmic rays is actually rather tricky. Unlike in an experiment at a

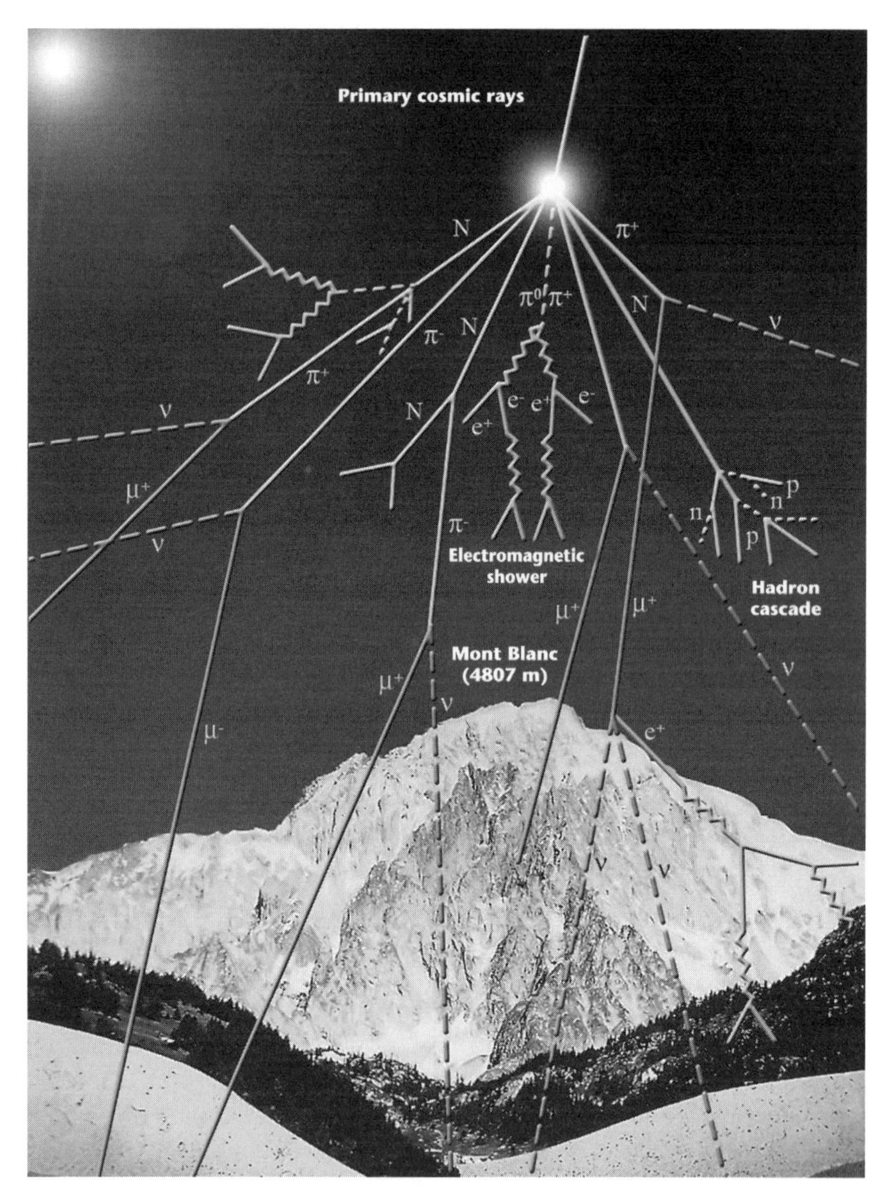

Figure 7.19. Protons from space hit air molecules, resulting in a shower of particles that pass through the atmosphere and hit the Earth's surface. Figure courtesy of CERN.

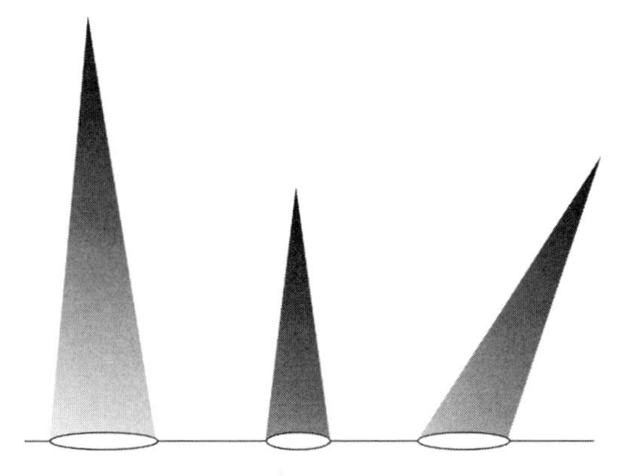

Figure 7.20. Three identical cosmic rays at different heights and angles of incidence. The relative darkness of the gray at the ellipse represents how much of the shower will be observed at ground level. This illustrates the need to know in great detail the height, angle, and other information about the evolution of cosmic ray shower.

particle collider, you actually have very little information. You can't measure the energy or identity of the particle from space prior to the collision. You can't put instruments around the spot in the atmosphere where it collided with the air molecule. You can't observe exactly how the debris from outer space travels through the atmosphere. The only thing you can do is construct a detector that sits more or less at sea level—10 to 20 miles (20 to 30 kilometers) from the collision itself. From this meager information, we try to figure out the energy of the particle from space and even the direction from where the particle came.

Figure 7.20 shows what the cosmic ray experimenter is up against. Three identical cosmic rays interact in the atmosphere at different altitudes and with different incoming angles. The density of particles in the cosmic ray shower is greatest near the collision point and dies off as the particles are slowed and stopped in the atmosphere. In the figure, this loss of particles is illustrated by the lightening of the shower in the cone. The only measurement occurs in the ellipse at ground level. The figure clearly shows that these three identical cosmic ray showers will have very different signatures in the ground-based detectors.

Using a single measurement to determine the energy of the highest energy cosmic rays requires a few things. First, you need a large detector, as a large shower can cover square miles of area at ground level. (Note the Auger detector covers 1,000 square miles [2,500 square kilometers] of the Argentinean desert, or an area about the size of Rhode Island.) Second, you need some method for determining both the direction and height of the start of the shower. This is usually done with certain timing techniques, but other methods are also used.

The third requirement—and one most relevant to the LHC—is a good understanding of just what happens when a very high-energy proton hits another one. Without a good model of this process, the study is all guesswork. The problem is that there has never been a way to test our predictions of how such high-energy protons interact with matter. The LHC is able to provide tests never before possible. While not strictly speaking part of the LHC's main mission, the LHCf experiment will make measurements that will help out a lot.

In this chapter, we have only mentioned the tiniest fraction of questions that have been and will continue to be explored by the various experiments arrayed around the LHC. Extra dimensions, supersymmetry, and quark structure are some of the main topics have begun to be studied, but, with approaching 10,000 experimental physicists involved, you can expect a veritable torrent of scientific results to come from these efforts. From the CMS experiment alone, more than three hundred papers have been published as of the summer of 2013, with many more on the way. The ATLAS publication record is comparable.

What will be the next big discovery? I have no idea. It may well be one of the topics mentioned here. Or, even more exciting, it may be something utterly unexpected; something that just hits us out of the blue. As they say, time will tell.

Although all these fascinating physics ideas are interesting to consider, it is only through experiment that we will know which one is right. The LHC is off to a good start, but there is much more to do. In the next and last chapter, we will recap what the future is expected to bring.

8

THE FUTURE IS BRIGHT!

The universe is an amazing place, full of dizzying complexity and breathtaking beauty. If you think about it just the tiniest amount, the phenomena we encounter as we make our way through our day can only be described as mystifying. Liquid water turns into steam as we brew our morning tea. It's misleading to say that we head out the door, as we actually have to open the door. Literally heading out the door will result in a bloody nose because it's somehow solid, rather than the gaseous air that we move through. What makes some things solid and others gas?

As we leave the house, we are warmed by the sun, a glowing ball of plasma located over 90 million miles away, heated to millions of degrees by miniscule atoms slamming together and binding themselves into heavier atoms, releasing energy in each interaction. We, ourselves, are mobile bags of chemicals—salt and water, calcium and iron—that somehow convert our morning muffin into the ability to go about our business. We live and we die and sometimes make babies along the way. Over the course of the days and years of our lives, the sky alternates between brightness and dark. Sometimes it's blue. Sometimes it's gray. Sometimes it's hot and other times it's cold. The more observant of us will see some points of lights march across the sky but maintain their position relative to each other, while other lights wander across the heavens in predictable and stately paths—the planets, the moon, comets, with an occasional fleeting shooting star to liven things up.

These are just a fraction of the phenomena mankind has encountered since our species branched off from the other great apes. For thousands, or even hundreds of thousands, of generations, our ancestors encountered this complex environment, learning the rules of our universe: which rules were important and which were not; which rules could help and which could kill. For many of our great, great, and many more greats grandparents, the curiosity was a practical one. And a good thing it was that it is practical, for their curiosity helped them exploit their environment so that they could survive and thrive, until now when ours is the dominant species on Earth, with power to live in every conceivable environment, indeed the ability to change the environment; to reshape the very surface of the planet; to defeat diseases that once ravaged millions.

The fact that our species has been able to evolve from tribes of bipedal hominids—living lives not markedly different from those of the wild chimpanzees and great apes of Africa—to a community that spans the globe with unprecedented power over nature stems from many things: our upright posture, our dexterous hands with opposing thumbs, our ability to vocalize in ways that allow us to communicate, and arguably the most important of all, our brains. This modest little organ has the ability to study the world and discern complex and subtle patterns. And, for a few of us, the study of these patterns has become an obsession.

During the course of millennia, some men and women have observed the world, trying to understand not the practical rules but the ultimate ones. They wanted to understand not simply that heated solid ice turns to liquid, while heated solid wood turns to ash, but *why* that happened. Once the fundamental rules were unveiled, our eyes would be opened, revealing what Einstein once called "God's thoughts," which are the basic, underlying reasons for why the universe is the way it is.

Over the generations, these deeper seekers have had had various names, from priests and shamans, to alchemists and natural philosophers, to our modern term of scientists. There are many kinds of scientists in the world nowadays, each studying the world, learning about this thing or that. However the ones who have held most true to humanity's heritage of understanding the universe at the deepest and most fundamental level are the physicists. Not just any kind of physicists but rather the particle physicists and cosmologists.

This is not to say that my colleagues and I are the only ones who study difficult and worthwhile questions. The study of the origins of consciousness is the province of the neurophysiologists, and understanding how life arose from collections of inanimate chemicals is the playground of the biologists. The biblically contentious question of the origin of mankind will only be resolved through the dutiful efforts of anthropologists and paleontologists. But, no matter how interesting these questions are (and they are fascinating indeed), the answers that are found will be bound by the strictures of biology or chemistry. The study of chemistry is governed by the behavior of atoms, which are nothing more than collections of protons, neutrons, and electrons, held together and guided by a handful of forces. A curious person might ask why protons and neutrons have the nature they do. This is the reason the Large Hadron Collider was built . . . to answer the deepest questions ever imagined: Why is the universe the way it is? Could it be different? If so, why this way and not others? What are the most basic rules of force and matter, of space and time—the rules from which all other rules are derived?

In chapter 2 we looked at our current understanding of the rules of the microrealm. These rules are ones of sublime simplicity. Two different kinds of particles (quarks and leptons) governed by four forces (gravity, electromagnetism, and the strong, and weak nuclear forces) are sufficient to build our entire universe, from you and me to galaxies and all of the visible cosmos. Even those thorny questions of the mind and the origins of life are bound by and will eventually be explained by these simple building blocks and governing rules, albeit in far more complex forms, probably involving emergent properties of these principles. Emergent properties are complex phenomena that originate from simpler ones, like the way that the simple rules governing how water freezes can lead to the astounding diversity in the types of snowflakes. To explain snowflakes, one need not invoke anything new beyond the law of electromagnetism that builds water molecules. Often the complexity arises from the interactions of many molecules and these more complex behaviors aren't easy to predict when considering just the simplest case.

One should not believe that the quarks and leptons and the four fundamental forces are the entire story. If they were, there would be no more questions to answer, and that is clearly not true. When the LHC was proposed, the question of the origin of mass of fundamental particles was not resolved. Strictly speaking, this state of affairs remains as of this writing, even though we are much closer to answering that question than we once were. Even if the original Higgs hypothesis described in chapter 6 is the final word, the Higgs hypothesis wasn't rooted in a deeper theory. It will require a better understanding of the rules of the universe to determine why the Higgs idea is right.

The LHC was built to make it possible for scientists to be able to explore, to discover, to learn. As we have seen in these pages, it was completed in 2008, designed to collide beams of protons together at the unprecedented energy of 14 trillion electron volts (TeV) and at nearly a billion times a second. Some overlooked engineering problems showed themselves during the last stages of the shakedown phase and the result was that the LHC didn't turn on until the spring of 2010, at half the design energy and with a beam brightness that was spectacularly below the original plan. Still, this energy was three and a half times higher than the previous record holder. The experiences of 2010 were used not only to understand how the new equipment would work but also to explore unfamiliar territory. In 2011, accelerator physicists grew comfortable with their new toy and delivered in one year half as much beam as the Fermilab Tevatron recorded over an entire decade. The writing was clearly on the wall, leading the

Tevatron to shut down for good in late September 2011. In 2012, the LHC took its undisputed place as the facility that will dominate the energy frontier for the next two decades or more. During this year, beam energy was raised from the 7 TeV of 2010 and 2011 to 8 TeV and the amount of beam delivered was four times the LHC saw in 2011.

As I write this, the LHC is in the middle of a planned shutdown period. During this time, the original engineering design flaw of the LHC will be rectified. The goal is to resume delivering power to the LHC in the fall of 2014, around the time this book gets in your hands. For the first few months, the accelerator scientists and operators will begin to learn how the machine actually operates at its new energy and with all the kinks ironed out of it. Any facility as big as the LHC will have idiosyncrasies that need to be understood. Presumably beam will be circulated, first at lower energy but then at higher and higher energies.

In the early spring of 2015, the plan is to return the LHC to full operations, this time at 13 or 14 TeV. The actual number will not be determined until the lead LHC scientist has listened to reports from his staff and considered the merits and dangers associated with the various possible operating points. He (or she, but it is currently a he) will present the options to the CERN director for a final decision.

In addition to raising the energy of the accelerator, another change will be to collide the beams every 25 nanoseconds. This was the original design, but during the 2010 to 2012 running period, the beams were colliding every 50 nanoseconds. If it is possible to make this change, this will make the data analysis easier, since each time the beams pass through one another, there will be half as many collisions (because the beam will be spread out more). This will greatly simplify the lives of the scientists who must sift through the debris of the collisions to figure out what message the data are telling us.

From perhaps the spring of 2015, the LHC is planned to run through the end of 2017. This will be followed by a yearlong shutdown for upgrades and then another running period of three years' duration. During this period, the beam brightness might double. After that, the future is a bit fuzzier. Some plans include significant improvements in beam brightness. No matter what happens, the existing detectors will be battered and ailing and need considerable refurbishing.

The broader picture is also fuzzy. Newer accelerators will be expensive and the world's economy is uncertain. Physicists have planned for a new facility called the International Linear Collider, or ILC. This accelerator will bring beams of electrons and positrons to high energy and collide them together. The proposed ILC is different from the LHC in that the electrons and positrons annihilate and all the energy is used in the collision. In contrast, the LHC collides protons, which are essentially bags of quarks and gluons. Because it is impossible to predict the energy carried by each quark or gluon, it is also impossible to predict the energy of each collision. If you want to dial in a specific collision energy, you use an electron/positron collider. As discoveries and measurements come out of the LHC, it will be clearer what energy the ILC should be built to deliver.

It's impossible to predict where the ILC will be built, indeed if it's built at all, but the smart money is currently betting on Japan. However, my crystal ball is a bit cloudy, and I am not predicting the state of the field a decade from now.

Still, the near-term future is very bright. The LHC will be the center of the energy frontier for the foreseeable future. Scientists who have followed a tradition that spans the millennia of needing to search for the answers to the ultimate questions will find their way to CERN.

I'll see you there . . .

Suggested Reading

General Information on Particle Physics and the Large Hadron Collider

No one book can explain such an interesting set of topics in enough detail to satisfy every reader. Frequently a reader might want to know a lot more about a topic mentioned only briefly. Luckily, there are many good books written for a lay audience on many physics topics, and readers can frequently find that deeper and more thorough explanation that they are wanting. In this section, I try to recommend some of the better books available on the subject matter covered in this book

It is difficult to organize such a list, as books often have multiple strengths. Thus I have chosen to list the books, with some commentary and a list showing which chapters the books overlap with.

For general information about the world we know, I recommend *The Particle Garden* by Gordon Kane. It is a short book that describes very clearly what we know. It is written by a theoretical physicist, so it is light on experimental details. (Chapters 2, 6, and 7)

For a more experimental treatment, my own *Understanding the Universe: From Quarks to the Cosmos* is a better choice. This book is much longer and covers the history of particle physics, our current understanding of the standard model, accelerators and detectors, current mysteries, and particle physics links to cosmology. The treatment in this book is aimed at a lay audience, but is at a slightly more detailed level than the present book. (Chapters 2 through 7)

I wrote another book about the LHC called *The Quantum Frontier: The Large Hadron Collider*, which told the story of the accelerator before it turned on. That earlier book has some noticeable commonalities with this one, but the book you're holding contains a lot more of the human element of the LHC story. Because the LHC hadn't started operations when the other book was completed, those stories hadn't been told yet. (All except chapter 5 and the discovery tale in chapter 6)

For a light and breezy treatment of the history of particle physics, interspersed with a discussion of the world as we currently understand it and culminating in a very short and non-technical discussion of the Higgs boson, try Leon Lederman and Dick Teresi's *The God Particle: If the Universe Is the Answer, What Is the Question?* Lederman's folksy demeanor and Teresi's professional writing background are apparent throughout. (Chapter 2)

Gordon Kane's *Supersymmetry* is a book written ostensibly for a lay audience on the topic of supersymmetry and walks a very fine line between a lay audience and a non-mathematical treatment for a very junior scholar. For any serious first exposure to the topic, this book is a must. The reader should be aware that Kane is an ardent proponent of supersymmetry, so there is some merit to critic's comments that the book is not perfectly balanced and it leaves the reader with the impression that the existence of supersymmetry in the world is a more foregone conclusion than it actually is. Dan Hooper's *Nature's Blueprint: Supersymmetry and the Search for a Unified Theory of Matter and Force* covers similar material and is really a good read. (Chapter 7)

For a discussion of the important role that symmetry plays in modern particle theories, the book *Symmetry and the Beautiful Universe*, by Leon Lederman and Christopher Hill is really quite nice. The idea of symmetry is sometimes daunting to the casual student of physics, and these authors do a good job of demystifying the topic. (Chapters 6 and 7)

For an accessible treatment about what we know that is somewhat more technical than what you've read here, try *Deep Down Things* by Bruce Schumm. The reader should be aware that Schumm's book does break the taboo of popular literature, by occasionally including an equation. But these equations are used as spice rather than as an obstacle to understanding and this choice will be welcome to all but the most math phobic. (Chapters 2, 6, and 7)

The Charm of Strange Quarks: Mysteries and Revolutions of Particle Physics, by Michael Barnett, Henry Muhry, and Helen Quinn is an unusual book. It covers the usual subjects, but the format is a mix of book, magazine, and textbook, with sidebars, column notes, and professionally drawn graphics. It has a vague similarity to the "X for Dummies" series (although it is entirely unrelated.) It also is one of the few books that have any treatment of detectors. (Chapters 2, 3, and 4)

Another unusual book is *The Particle Odyssey: A Journey to the Heart of Matter*, by Frank Close, Michael Marten, and Christine Sutton. This book can be described as a "coffee table book," with extensive color photographs. It is a photo montage that includes history and future, even including some photographs of LHC prototypes. For those who need to see something to understand it, this is a very valuable book.

There are readers who like the stories of the history and the personalities as much as the physics. For those readers, Martinus Veltman's *Facts and Mysteries in Elementary Particle Physics* is a good choice. In addition to the usual descriptions of the physics we know, Veltman intersperses the text with one-page asides describing many of the colorful characters who have helped us understand our universe. As a Nobelist himself, Veltman is personally acquainted with many of these people and so many of his anecdotes have a firsthand flavor. Veltman does mention accelerators and detectors, but the cursory treatment reflects his own high achievement as a first-rate theoretical mind. (Chapters 2, 6, and 7)

While the history of the discoveries of particle physics in the twentieth century is not a focus of this book, for a reader who is interested in the subject, I recommend the brilliantly written *The Second Creation*, by Robert Crease and Charles Mann.

An interesting book that describes the discovery of the Z and W bosons and gives a real sense of the excitement and competition that goes along with a Nobel-bound discovery is *Nobel Dreams* by Gary Taubes.

For a person who is interested in the history of CERN, it is hard to compete with *History of CERN* by A. Hermann et al. (volumes 1 and 2) and J. Krige (volume 3). These books are quite expensive and rare, so an interlibrary loan is your best bet for these.

The astute reader will note that most of the suggested reading is related to chapters 2, 6, and 7, which is to say what we know and what our theories are looking for. Chapters 3 and 4, which describe the accelerator and detector principles, as well as details of the LHC complex, are uncommon. Partially this is because many books emphasize the theoretical side of things. I expect that this will change as time goes on. It does also reflect an attitude among some that these are merely tools and not as interesting as the discoveries they make possible. However, the history of science has always been an interplay between the discoveries and the equipment. It is impossible to fully appreciate why we believe the things we do if we don't understand the evidence. And one can never understand the evidence without an appreciation of the tools.

For the more technically minded I recommend the journal article "General-Purpose Detectors for the Large Hadron Collider" by Daniel Froidevaux and Paris Sphicas in *Annual Reviews of Nuclear & Particle Science*, volume 56, pages 375–440, published in 2006. Note that, as this is a journal article intended for other particle physicists, it is definitely *not* easy reading.

Web Sites

Web sites are always a dangerous thing to publish, because the World Wide Web is a fluid place, and things change rapidly. Some sites are likely to exist for some time and would be helpful for the avid reader. They are:

Don Lincoln Facebook: http://www.facebook.com/Dr.Don.Lincoln
The CERN home Web site: http://www.cern.ch/
The press office for CERN: http://press.web.cern.ch/press/
The ATLAS experiment: http://atlas.ch/
The CMS experiment: http://cms.cern.ch/
The LHCb experiment: http://lhcb.web.cern.ch/lhcb/
The ALICE experiment: http://aliceinfo.cern.ch/Public/

Interesting particle physics news and images from across the world:

http://www.interactions.org/cms/

Additional Suggested Reading Made Available after the LHC Startup

Once the LHC began its operational phase, it is inevitable that a number of new books entered the market that discuss the equipment and the physics program. The most introductory book is *Collider* by Paul Halpern. This book is more historical than my book *The Quantum Frontier* and is focused on the development of the colliding accelerator. The two books cover quite different material and are complementary. (Chapter 3)

There are three other books available about the LHC. All of these books are at a fairly advanced level. These are *Perspectives on LHC Physics* by Gordon Kane, which discusses the physics goals and expected discovery capabilities of the LHC. In addition, there is a definitive book edited by Lyndon Evans, the lead scientist on the LHC project. This book is called *The Large Hadron Collider: A Marvel of Technology*. This book is rather technical but is important for the serious lay reader. The final book currently available was written by Dan Green and is called *At the Leading Edge*. This book is an anthology and covers the main subsystems of the CMS and ATLAS detectors. Again, this book is for someone interested in a more technical treatment. (Chapters 3 and 4)

The final report of the committee investigating the LHC incident of September 2008 is *Report of the Task Force on the Incident of 29 September 2008 at the LHC*. This is LHC Project Report 1168 and can be found at http://cdsweb.cern.ch/record/1168025/files/LHC-PROJECT-REPORT-1168.pdf.

An unusual information source is the 2009 article "The Black Hole Case: The Injunction against the End of the World," *Tennessee Law Review*, volume 76, no. 4, pages 819–908, written by Eric Johnson. This is a thorough article describing the legal issues that could be considered in a hypothetical case to keep the LHC from operating on the basis of worries about the creation of black holes. It is a very accessible resource on the various points and counterpoints that have been raised on the issue. It does not do a good job describing the probability of the scenarios, but it presents the physics concepts in a very accessible way. Also available at: http://arxiv.org/abs/0912.5480.

Ian Sample's *Massive* is an excellent compilation of the history of the Higgs boson. It focuses less on the physics and more on the people who made it happen. In this book, you'll find that calling it the "Higgs boson" definitely plays down the contribution of quite a few clever scientists. (Chapter 6)

Sean Carroll's book *The Particle at the End of the Universe* and Jim Baggott's *Higgs* were written after the discovery of the Higgs boson and thus they tell parts of the story I've told here. Neither of them are LHC experimenters, so they tell it from an outsider perspective. Carroll's book is an easy read. Baggott's book is somewhat misnamed, as it concentrates on symmetries more than the Higgs boson. Baggott's book also has some physics mistakes that will annoy particle physicists but do not detract from his story. (Chapters 2, 5, 6, and 7)

There are two books that I can't say I have read, which have come out since my original LHC book, *The Quantum Frontier*, was published. They are Gian Francesco Giudice's *A Zeptospace Odyssey* and Amir Aczel's *Discovering the Higgs Boson*. I have thumbed through both, and they are on my list of books to read when I get time.

If you're interested in something for your children, there is Bonnie Juettner-Fernandes' *Large Hadron Collider*, which is written at a level appropriate for 8- to 10-year-olds. There are a lot of pictures, some trivia, and interesting physics. While I did not write it, I was the scientific consultant for the book.

Index